Riddles in you...

Second Edition

Riddles in your Teacup

Fun with Everyday Scientific Puzzles

Second Edition

Partha Ghose
S N Bose National Centre for
Basic Sciences, Calcutta, India

Dipankar Home
Bose Institute,
Calcutta, India

> **" With faith in the unsearchable riches of creation and the untried fertility of those fresh minds into which these riches will continue to be poured, I wish you success. "**

JAMES CLERK MAXWELL

at the inauguration of the Cavendish Laboratory, Cambridge.

Contents

Foreword

The fascination of science derives in large part from the fun of problem solving. I often compare the scientific method with doing crossword puzzles. Nature provides us with clues, often cryptical in form, and it requires much insight and ingenuity to 'solve' these clues. But such is the wonderful coherence of nature that each "solution"—a law of physics, a new phenomenon, a fundamental principle—beautifully interweaves with others to make a consistent pattern.

The world is full of surprises. Some take the form of baffling natural phenomena, such as the ability of trees to draw water many metres above the ground; others relate to domestic oddities, like the singing of a kettle. Mostly these puzzles can be solved by a careful application of simple, high-school science, though some require more subtle and advanced concepts. But with each solution there is a sense of the triumph of reason and rationality over mystery. Science works!

The great American physicist Richard Feynman once remarked that in dealing with puzzling and complicated physical processes, if you ask a clear question nature will provide a clear answer. The trick of being a good scientist is to know which question to ask. Once you hit on the right way to think about something, the solution invariably pops up effortlessly.

Science teachers are often accused of taking an unnecessarily formal and abstract approach to their subjects, and ignoring the relevance of science to everyday life. It is therefore especially welcome to find a book that focuses on the delightful oddities that surround us in the home and our immediate environment. Partha

Ghose and Dipankar Home have painstakingly collected a wide range of scientific curiosities, and given us simple explanations.

Although the book will appeal especially to the young, all scientists are young at heart, and there is much here to amuse and inform everyone. As I read through the examples, I continually found myself saying: "Yes! I've always wondered about that!" Well, this book gave me the chance to find out the answers. Its lively tone and easy-to-read style make it an ideal companion book for the more formal and traditional texts of the classroom.

The most useful aspect of this collection is the way in which the problems chosen encourage us to think creatively about the world. So often, science is presented as a dry set of conventional procedures, whereby contrived experimental arrangements lead to precise and "correct" answers. But science is not just something that happens in laboratories. How refreshing to see scientific principles put to work in the world about us, which is often complex and messy, and where a satisfactory explanation may involve surprising or counter-intuitive ideas.

The authors are to be congratulated on presenting science with a human face. Readers will find that as they peruse the topics, many other examples will come to mind, curious little things about the world that intrigue or mystify them. If this book encourages them to seek out their own explanations then it will have played a valuable part in the education process.

Paul Davies, Adelaide, Australia

Preface to the Second Edition

*66The most beautiful experience
we can have is the mysterious.
It is the fundamental emotion
which stands at the cradle of
true art and true science.
Whoever does not know it and
can no longer wonder, no longer
marvel, is as good as dead, and
his eyes are dimmed. 99*

ALBERT EINSTEIN

One of our greatest pleasures over the last few years has been interacting with the young and "playing" with physics: trying to understand commonplace phenomena in terms of basic physical principles and delighting in their profundity, generality and their subtle interplay with reality.

Our familiarity with natural phenomena and the ordinary things that happen every day around us robs them of their mystery and makes them seem obvious to us; we stop wondering at them. Yet more often than not they conceal delectable surprises and puzzles. Identifying and grappling with them is a fascinating quest.

This book has grown out of our regular columns in Indian popular science magazines and weekly newspaper columns. We owe a lot to our enthusiastic readers who have helped not only with answers but also with problems. Some of these problems appear in this book. The initial impetus, of course, came from the

Indian television programme "Quest" in which one of us had the privilege to participate for a while. The first edition, published in 1990, has been *revised* and *expanded* considerably for this new edition.

We have arranged the book into several sections, not according to the conventional partition of physics into heat, light, sound and so on, but according to whether we face the puzzles in and around our kitchen, out there in nature, on the playground, watching a movie, or reading a novel. We find this way of classifying more interesting and natural. The last section contains a few riddles that, to the best of our knowledge, either still remain unsolved or whose solutions are not that straightforward.

The book is intended for students and lay persons with a high-school background. We have tried to keep the answers simple and intuitive. This has meant that we have not always been able to be comprehensive or sufficiently penetrative. This book is primarily for enjoyment and we hope some of our readers will be stimulated to look for more technical answers elsewhere such as in the books cited in the Acknowledgments.

We urge you to go through this book critically. And if, as a result, you are able to solve one or two of the open problems or notice new ones or have anything to say about the answers we have given, do write to us, care of the publishers. We would love to hear from you.

Partha Ghose and Dipankar Home
Calcutta, India

Acknowledgments and Bibliography

We enjoyed collaborating with Suparno Chaudhuri who did the illustrations for the first edition, which have also been used in the second edition.

In preparing this *revised* and *enlarged* edition we had the benefit of perceptive suggestions and helpful criticisms from Paul Davies, John Gribbin, Nigel Henbest, Neal Marriott, Andrew Robinson, Euan Squires, C S Unnikrishnan and Andrew Whitaker. We thank Paul Davies for finding time to write a stimulating Foreword to this edition. We are also indebted to the books listed below which have helped us considerably. Readers who are interested in more scientific details and references concerning a number of problems occurring in our book will find them in the following books: (a) M. Minnaert, *The Nature of Light and Colour in the Open Air* (Dover, 1954); (b) J. Walker, *Flying Circus of Physics* (Wiley, 1977); (c) R. Greenler, *Rainbows, Halos, and Glories* (Cambridge University Press, 1980); (d) C. F. Bohren, *Clouds in a Glass of Beer* (Wiley, 1987); and (e) C. F. Bohren, *What Light Through Yonder Window Breaks?* (Wiley, 1991).

The photographs of Schrödinger, Dirac, Feynman and Bohr and Pauli appear courtesy of AIP Neils Bohr Library and that of Einstein appears courtesy of ETH Bibliothek. Thanks are due to Anindita Home for help in preparing the *revised* manuscript for this edition.

Questions

Kettle Croon
Physics around the Kitchen

❝All our knowledge brings us nearer to our ignorance. ❞

T S Eliot

❝It isn't that they can't see the solution. It is that they can't see the problem. ❞

G K Chesterton

Kettle Croon

We are all familiar with the hissing sound (called the "singing" of the kettle) that starts a few moments after the kettle is put on the fire to boil water. This sound gradually increases and then suddenly drops when the water starts to boil. In fact, we know from the sudden drop of the sound that the water is ready, boiling. Have you ever wondered what causes the kettle to "sing"?

Spoon in a Teacup

One often puts a metal spoon into the china cup before pouring hot tea into it. Why? Which is safer to use, a thin-walled cup or a thick-walled one?

Einstein in your Teacup

Erwin Schrödinger was an eminent physicist who discovered the fundamental equation of quantum mechanics which describes the behaviour of atomic and sub-atomic entities. Schrödinger's wife remembered Einstein every time she poured her tea. This is because it was Einstein who first explained to her and to her husband why wet tea leaves (which are heavier than the liquid) always collect at the centre of the bottom of a cup when the tea is rotated by a spoon for a while and then allowed to settle. This is what Schrödinger wrote to Einstein on 23 April 1926 (reprinted in *Letters on Wave Mechanics*, edited by K Przibram, Philosophical

Erwin Schrödinger (1887–1961) was born and educated in Vienna, Austria. Until the age of 11 he was taught at home, and his father encouraged his interest in nature with a microscope and other equipment. In 1926 he discovered the central equation of quantum mechanics for which he shared the 1933 Nobel Prize for physics with Paul Dirac. Photograph courtesy of AIP Neils Bohr Library.

Library, p 27): "It just happens that my wife had asked me about the 'teacup phenomenon' a few days earlier, but I did not know a rational explanation. She says that she will never stir her tea again without thinking of you."

Next time you have tea, turn it with your spoon and notice where the leaves settle. Why do you think the leaves settle at the centre and not get pushed to the walls by the centrifugal effect?

Paul Dirac (1902–1984) was born and educated in Bristol, England, before going to Cambridge University to do research. He shared the 1933 Nobel Prize for physics with Schrödinger for the development of quantum mechanics. Dirac was an extremely original thinker but notoriously reticent. This may explain why he is still relatively unknown to the general public. Dirac's friend Peter Kapitza, the Russian physicist, once gave him an English translation of Dostoevski's classic Crime and Punishment. *Later when Kapitza asked him how did he like the book Dirac replied in his characteristically succinct way: " It is nice, but in one of the chapters the author made a mistake. He describes the Sun as rising twice on the same day". Some of the readers of this present book may wish to try to find out in which chapter this occurred.* Photograph *courtesy of AIP Neils Bohr Library.*

A Hole in a Tea Pot
Why is a small hole usually made on the lid of a tea pot?

The Teetotaller's Dilemma
Some like to pour milk first and then tea, others prefer to add milk to the tea. Is there any difference between the two?

Fire without Hazard
Why doesn't the whole gas cylinder catch fire when the burner is ignited?

The inner Core
When one makes ice cubes in a refrigerator, one usually finds that the outer part of the cubes is transparent whereas the inner core is opaque. Why?

An Apple a Day
Why does the cut surface of an apple turn brownish after some time?

Ovens with a Difference
Microwave ovens are now quite common in kitchens. Do you know how they work?

Don't Lick an Ice Tray

Have you ever tried to hold a really cold frosted ice tray? If you have, you must have noticed that your fingers tend to stick to the tray. Why? Don't ever try to lick the tray—it will be a very painful experience!

From Fermi to the Frying Pan

The famous Italian physicist Enrico Fermi once asked a student during an examination: "The boiling point of olive oil is higher than the melting point of tin. Explain how it is then possible to fry food in olive oil in a pan". (Italian saucepans are wholly made of tinned copper.) What is the answer?

Coiling Chocolate
The coiling of thick molten chocolate as it is poured onto a plate or a slab of ice-cream must have struck you as odd. What on earth makes it coil?

Leaping Liquid
The nuisance caused by milk spilling over when boiled is well known. One has to keep a constant watch and stir the milk to prevent it spilling. Why does milk have this peculiar property?

Soup Swirl

Next time you have a thick soup at lunch, or prepare a paste of starch, give it a good swirl, lift your spoon and watch for a few seconds. You will notice that just before the turning stops, its direction reverses momentarily. This phenomenon illustrates an important characteristic of real fluids, namely, . . . what?

Honey of a Problem

Pour out honey gently from a jar. If you intercept the thin stream of falling honey with a knife, you will see that the honey above the knife shrinks back and disappears into the jar. Don't pour out the honey too quickly; let it just trickle down. What do you think causes this "antigravity" effect?

Our Daily Bread

**_❝One of the strongest motives
leading to art and science is a
flight from everyday life with its
painful coarseness and bleak
tediousness, from the chains of
ever-changing personal wish. ❞_**

ALBERT EINSTEIN

**_❝The essential point in science is
not a complicated mathematical
formalism or ritualised
experimentation. Rather the
heart of science is a kind of
shrewd honesty that springs
from really wanting to know
what the hell is going on. ❞_**

SAUL PAUL SIRAG

Have a Drink

When we drink, we bring the glass or cup containing the liquid near our lips and suck in the liquid. What makes liquid rush up into our mouth? Take a bottle of some drink, cover its mouth with your lips and try to suck in the drink without inverting the bottle above your mouth. What happens?

Soap and Dirt

How does soap help remove dirt from our bodies and clothes? Any idea?

Funny Funnel

You must have noticed while pouring a liquid into a bottle through a funnel that you have to lift the funnel from time to time when the liquid collects in the funnel and does not flow down. Do you know why?

Blow Out

Who hasn't blown out a candle or watched one being blown out by a gust of wind? Even such a commonplace phenomenon is however quite baffling. Why should a candle be blown out in spite of a supply of more air (containing oxygen which helps burning)?

Iron it Softly

It is a common practice to sprinkle some water on a starched cloth before pressing it with a hot iron. Why does it help to sprinkle water and then use a hot iron?

Fire! Fire!

Whenever there is a fire we wish to extinguish, we think of water. The fire brigade uses water to put out big fires; in India one sprinkles water on the kerosene stove after cooking is over. What makes water an effective fire extinguisher?

Ice Fumes

Have you noticed that when exposed to air, a large slab of ice appears to give out fumes? What are these fumes and why do they form?

Coasting Along

Why does a coaster tend to stick to a wet-bottomed glass when the glass is lifted?

A Touch of Chill

At room temperature, particularly during winter, a metal chair feels much cooler than a wooden or plastic chair. Why?

Tractors and Farmers

A heavy crawler tractor is able to operate on soft, muddy ground but the farmer's feet sink. Why?

Blinding Light

We are annoyed when cars coming from the opposite direction have their headlights on, because the bright light dazzles our eyes. Also, when there is a power cut, for a while we are unable to see anything. Then gradually our eyes get adjusted and we are able to discern faintly the objects around us. Why do our eyes react to light the way they do?

Rest in a Hammock
Why is it pleasant to lie in a hammock though the pieces of rope that go to make it are by no means soft? Why is it pleasanter to sit on a wooden chair rather than on a flat-topped stool?

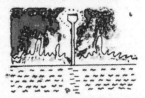

Long and Broken
The image of a street lamp on a lake or pond often appears elongated and broken; a very common sight. Do you know why?

Boot Polish
A friend's son was polishing his shoes the other day. Neither the sticky polish nor the brush had anything that he could connect with the shine of the shoes. He was puzzled. Can you help?

Tear a Paper

When you tear up a piece of paper, you can hear a characteristic sound. Notice that the quicker you tear it, the higher is the pitch of the sound. Any idea why?

Woof, It's Cold!

If you have been to any hill-resort you must have noticed that it's cooler up there, yet the Sun is fiercer on the skin (It's easier to tan). Why is it cooler at higher altitudes than at the sea level, even though you are several thousand feet nearer the sun?

Foggy Mirror

Many of you must be familiar with the fogging of the mirror in the bathroom after a hot shower. The fogging of car windscreens during heavy showers is also a familiar sight. There is a simple way of avoiding this kind of fogging. Do you know what it is and how it works?

Roll a Coin

Place a slim coin vertically on its edge on a table. It tends to fall on its side. Now give it a push—it rolls forward steadily for a while without toppling. Why?

Snoring Away

Why do people snore?

Night Lends Clarity
Why is it that far-off radio stations are heard clearly at night but not during day time?

Perfumes are Airborne
How does the smell of perfume spread all over the room even when the air is still?

The Yellow Fog
Fog lights are usually yellow. Why?

The Painkiller Bottle
How does a hot water bottle relieve muscular pain?

Squeaky Chalks
Why does a chalk make a squeaky sound on the blackboard when not held at the correct angle?

Play Time

❝We dance around a ring and suppose. But the secret sits in the middle and knows. ❞

ROBERT FROST

❝It is better to debate a question without settling it than to settle a question without debating it. ❞

JOSEPH JOUBERT

Raman's Billiard Ball Problem

The Indian Nobel laureate physicist Sir C V Raman was passionately curious about everything that happened around him, including the sharp click that is heard when two billiard balls collide. Did you ever suspect that even such a simple and common phenomenon might involve unexpected subtleties of physics? Raman made a careful study of these clicks and arrived at some interesting conclusions. For example, he found that the intensity of the click varied with the direction around the billiard table. Can you guess in which direction it is a maximum and why? It might be well worth your while to do the experiment yourself and find out. If you cannot find billiard balls, try with marbles.

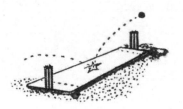

Play Cricket

A cricket ball often moves faster after pitching on a smooth wicket. What do you think is the reason?

Top Spin

Tennis and table-tennis players often use "top spin", which makes the ball dip and land earlier than expected. Why does the ball's spin (about the diameter perpendicular to its direction of motion) make it dip?

Follow Shots

In a game of snooker or billiards, one often sees "follow shots" in which the cue ball for a time follows the ball (of exactly the same mass) which it hits, even when the ball which it hits has picked up full speed. This seems to violate the principle of conservation of energy. How would you explain such follow shots?

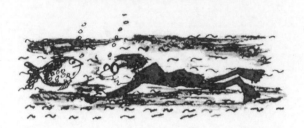

Swimming Underwater

Have you noticed that when swimming underwater, you can see much better if you wear goggles? Why?

Ride along

A cycle at rest cannot be made to stand on the ground, but given a rolling motion, it does not fall. What do you think is the reason?

Pole Vaulting

How does a pole vaulter gain such extraordinary heights?

Sleek and Swift
You must have noticed that racing cyclists and sprinters usually wear tight clothes and caps over their short hair. Why?

Cyclopean Vision
Why do we find it easier to aim with one eye?

Grand Jete
One of the most graceful movements of a ballet dancer is a "grand jete" in which the dancer takes a leap into the air and appears to glide parallel to the ground for a while. How is that done?

Soaring High
Why does a high jumper need to run up to a jump?

The Juggler's Trick
When one throws an object like a hat into the air, it wobbles and comes back, inverted or sideways or upright. How then does a juggler manage to catch a falling hat on his head every time?

Flow, Fluid Flow!

❝So long I have seen only with my eyes, now I want to see through my intellect".

RABINDRANATH TAGORE

❝The chief distinguishing mark of a scientifically oriented person is that he/she experiences physical discomfort at incomprehension and is not satisfied with an analgesic solution to a problem which merely relieves the ache of incomprehension without curing it".

PETER MEDAWAR

Feat of Flying
How does an aeroplane gain lift?

Of Birds and Aeroplanes
Is there any difference between how birds and aeroplanes fly?

Smoky Swirls
You must have noticed that when there is no breeze or draught, smoke from a cigarette resting on an ashtray rises steadily and smoothly up to a point and then suddenly breaks into swirls. Why?

The Fluttering Flag
The fluttering of a flag in the wind is one of the most common sights. Yet, how many of us have ever bothered to ask, why? Do you know what causes a flag to flutter in the wind?

Pour a Liquid

When you pour fruit juice or milk or any such liquid gently from a container, why does it tend to run down the side and not drop straight off from the lip? What factors determine how far down it adheres to the side of the container?

The Tapering Stream

Turn on a tap and watch the steady and smooth stream of water fall. You will notice that the stream narrows as it falls. Is there a force squeezing it together?

Expanding Smoke Rings

Have you seen veteran smokers puff out smoke rings? These rings are technically called vortex rings. They are remarkably stable in still air and can travel considerable distances without distortion. If such rings are directed towards a wall, it is found that they expand as they approach the wall. Why does the proximity of a wall make them expand?

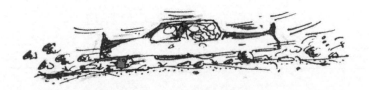

The Puzzling Balloons

A friend was once travelling in a car with his family, carrying helium-filled balloons. He noticed that whenever he accelerated the car, the balloons surged forward and crowded around his shoulders! Every time he put on the brakes, the balloons moved backwards and pressed against the rear window! Why did the balloons behave in such a crazy way?

An Anti-Gravity Effect

If you dip a capillary tube (a tube with a very fine bore) into a liquid, the liquid rises inside the tube. This is the mechanism that works in blotting papers which consist of fine capillary tubes. If you keep one corner of a towel dipped in water, gradually a large portion of the towel gets wet. Again it is capillarity in action; a towel has thousands of fine capillary cotton tubes through which water can rise. The question is: What is the source of energy that makes the water rise?

Through the Palm, Strangely!

❝It is a capital mistake to theorize before one has data. Insensibly one begins to twist facts to suit theories, instead of theories to suit facts .... The difficulty is to detach the framework of fact—of absolute undeniable fact—from the embellishments of theorists. ❞

Sherlock Holmes

❝There is no excellent beauty that hath not some strangeness in proportion. ❞

Francis Bacon

Through the Palm, Strangely

Roll a piece of paper into a tube. Hold it with one hand, say, the left hand, and look at a distant object through it with your left eye, keeping your right eye closed. Now, bring your right palm in front of your right eye so as to touch the tube, and then, open your right eye as well. You will see the distant object clearly through the hole in your right palm! (Both hands should be about 15–20 centimetres from your eyes.) It's fantastic, isn't it? How do you explain it?

It Does Not Pour Out

It is usually stated in text books that if you fill a glass completely with water, cover its mouth with a stiff card and invert it, the card sticks to the mouth of the glass and does not drop. Actually you will find that you do not need to fill the glass completely. Just pour some water into it, cover its mouth with a stiff card and invert it, and the card will stay in its place. What keeps it stuck to the glass?

Blow Hot, Blow Cold
Open your mouth and blow on to your palm—you will feel the warmth of the air coming out from your mouth. Now purse your lips and blow. This time you will clearly feel the difference—the air is cooler. Why?

Through a Glass Darkly
The details of a hand pressed against the frosted shower door are more distinct than those of the distant body. Why?

Incomprehensible Whispers
Have you noticed that sometimes you could hardly hear your friend's whisper when he/she was turned away, even if the whisper was as loud as the normal voice? Why?

Falling Cats
Cats are as nonchalant about heights as most of us are frightened—they regularly manage to survive falls from heights that would kill any person. How?

Ice in a Scarf
Take two chunks of ice, wrap one in a woollen scarf and leave the other open. You will notice that the one which is left in the open melts first. Not only does the scarf not give any warmth to the ice wrapped in it, it actually seems to help the ice to stay cool. How?

Dropping a Bottle
Imagine you are travelling in a car. You have a glass bottle in your hand. In which direction relative to the moving car should you throw it to minimise the danger of its breaking on hitting the ground?

The Burning Flame
Next time you carry a candle or a burning matchstick, notice that the flame is initially deflected backwards. Which way will it deflect if you carry it in a case?

Taper Caper
It usually takes some time to light a candle. But when a burning candle is extinguished and a burning splint is brought near it, the candle catches fire immediately. Why?

Wet a Brush

Take a paint brush. If you wish to make the hairs cling together, you would normally wet them. However, hold the brush inside water—the hairs do not cling at all. Why is it necessary to take the brush out of the water to make the hairs cling together?

Tyger! Tyger! Burning Bright

You must have noticed that the eyes of a cat shine brightly at night even when very faint light falls on them. This does not happen, for example, with human eyes. What makes a cat see better than us in the dark?

Hum with your TV

Philip C Williams observed (*Nature*, Volume 239, p 407, 1972) that humming at a certain pitch while watching television from a distance caused horizontal lines to appear on the television screen, which were visible only to the person who was humming. These lines could be made to remain stationary or move up or down by altering the humming pitch. Isn't that queer?

Play on a Ship

Two friends are playing with a ball on board a ship moving at a steady speed. One is standing nearer the aft and the other nearer the bows. Does one of them find it easier to throw the ball to his partner? (Ignore wind effects.)

No Spilling Over

Take an ice cube and float it on a brimful of water in a glass. When the entire cube melts, you will see that the water does not spill over. Why? This problem was made famous by George Gamow who claimed that he had put this question to a number of celebrated physicists and got conflicting answers.

To Catch a Card

Here's a trick you can try on your friends. Take a stiff card, like a picture postcard or a visiting card. Hand this to your friend. Ask her to hold it and, with her other hand, make a pincer. Then tell her to drop the card and catch it with the pincered fingers. Let her repeat the "drop and catch" sequence as many times as she wants to—she will catch the card every time. Now, tell her that if you were dropping the card, she would never catch it . . . . Try it. You'll win every time. Why?

The Eclipse of Superstition

There is a popular belief that solar rays are harmful during a solar eclipse. Is it true?

The Invisible Silver Thread

Though mercury is silvery white, why does it appear as a hardly visible black thread in a thermometer?

Which is Heavier?

Take two identical glasses filled to the brim with water, but one having a piece of wood floating on it. Which one is heavier?

Tearing Wet Paper

It is a common experience that it is much easier to tear wet paper than dry paper. Have you ever wondered why?

The Jumping Draught

Arrange a few identical draughts or coins in a straight line so that the neighbouring draughts or coins touch. Hold the first draught lightly with your fingers and strike it sharply on its edge with a ruler. You will see that the last draught or coin will jump away, leaving the rest in their places. Why?

Weigh a Stone in Water

Place a glass of water and a stone on one pan of a balance and balance them with weights. Then drop the stone into the water in the glass. What happens to the balance ? And why ?

Comb your Hair

When your hair is completely dry, you can try the following experiment. Take a small plastic comb and comb your hair or rub the comb with a piece of flannel. Then go near a tap and turn it on gently so that the water just trickles out. Hold the comb near the water. You will find that the trickle becomes a steady stream and is deflected by the comb ! Why? (The experiment works best in dry weather conditions.)

Weigh Yourself

Next time you weigh yourself, notice what happens when you bend forward. You will find that while bending forward, you seem to lose weight! Try another thing. Lift one of your arms quickly. This time you will find that while lifting your arm, you seem to gain weight! Why?

Darting Pepper

Take some water in a glass and sprinkle some pepper on it. Now rub a fingertip on a detergent soap and touch the water surface. You will be amazed to see how instantly the pepper particles will fly away from the spot in all directions. What makes them do that?

The Puzzling Rubber Band

Take a thick rubber band, stretch it quickly and hold it against your forehead—you will feel it distinctly warm! This is contrary to what one would normally expect. Remember that quick expansion of a gas usually cools it. Why does the rubber band behave in a contrary way?

*Fact and Fiction — In Movies and
Novels*

**❝Believe nothing, no matter where
you read it, or who said it, no
matter if I have said it, unless it
agrees with your own reason
and your own common sense. ❞**

Gautama Buddha

**❝Anyone who conducts an
argument by appealing to
authority is not using his
intelligence, he is just using his
memory.❞**

Leonardo da Vinci

Not With a Bang But a Whimper

One is familiar with silent gun shots from crime movies such as, for instance, Alfred Hitchcock's classic *North by North-West*. How does a silencer fitted to a gun function?

Fahrenheit 451

The title of Francois Truffaut's famous film *Fahrenheit 451* is related to the fact that one can boil water in an uncovered paper pot without burning the paper. Can you see the connection?

Wait Until Dark

Television news pictures taken in the dark of the night are now frequently shown. How is it possible to take such good quality pictures at night without using any additional light source?

An Oscar-Winning Problem

Those of you who have seen Sir Richard Attenborough's Oscar-winning movie *Gandhi* will recall a touching scene in the film where Gandhi gives his cotton wrapper to a poor woman. Gandhi takes out his wrapper, gathers it into a bundle and throws it into the river. The wrapper gradually stretches out beautifully on the water as it floats towards the poor woman. Why does the crumpled wrapper stretch itself out on the water?

The Invisible Man

H G Wells created the invisible man in his widely known story by the following trick: he made the refractive index of the invisible man exactly the same as that of air. So light rays simply passed through him without reflection or refraction. There is however a catch, a scientific fallacy involved in this conception. Can you figure out what it is?

Hiccupping Charlie

If you have seen *City Lights* you surely remember Charlie Chaplin going through a hilarious sequence of hiccups? What causes a hiccup?

The Humming Wires

Besides Charlie Chaplin, Satyajit Ray is the only other film director to have been awarded an honorary doctorate from Oxford University for his outstanding contributions to cinema. *Pather Panchali (Song of the Road)* is one of his most widely acclaimed films. In an enchanting sequence in this film, two children (Durga and her brother Apu) run around in a field, listening with wide-eyed wonder to the humming of the telegraph wires in the wind. Peter Sellers once wrote of this sequence: "It was so beautiful I could cry". Why do telegraph wires hum in the wind?

Can Lightning Magnetise a Sword?

There is a detective story (*The Royal Bengal Mystery*) by Satyajit Ray in which the detective solves the mystery by arguing that the suspected victim was not murdered but was struck by lightning. He used the following clue: the iron sword held by the man had been magnetised. There was also circumstantial evidence that lightning had struck the neighbouring area. Do you think it is possible for lightning to magnetise an iron sword?

The Ben Hur Chariot Race

Have you seen the classic film *Ben Hur*? Do you remember the spectacular chariot race sequence? If you do, you would recall that after the chariots picked up a certain speed, the wheels appeared to turn slowly in the reverse direction. The same thing also happens with rapidly rotating fans. Do you know why?

Doctor Zhivago

The "sweet, mellow" mood evoked by falling leaves during autumn has been used in creating memorable sequences in a number of films like in David Lean's *Doctor Zhivago*. Why do leaves turn red and gold and fall in autumn?

The Green Flash

The film version of Jules Verne's romantic novel *Le Rayon Vert* refers to a curious phenomenon that has been observed by many. Sometimes a green rim can be seen for a few seconds on top of the setting (or rising) sun. According to an old Scottish legend, anyone who has seen the "green flash" will never err again in matters of love. In the Isle of Man it is called "living light". Any idea what causes it?

The Murmuring Brook — Mysteries of Nature

❝Nature! Out of the simplest matter it creates most diverse things, without the slightest effort, with the greatest perfection, and on everything it casts a sort of fine veil. Each of its creations has its own essence, each phenomenon has a separate concept, but everything is a single whole.❞

GOETHE

Richard Feynman had a friend who was
an artist. He would often tell Feynman: "I, as an artist,
can see how beautiful a flower is. But you, as a scientist,
take it all apart and it becomes dull". Feynman disagreed.
He says in *What do you care what other people think?*
(Unwin Hyman, London, 1988): "I can imagine the cells
inside, which also have a beauty. There's beauty not just
at the dimension of one centimetre; there's also beauty at
a smaller dimension. There are the complicated actions of
the cells, and other processes. The fact that colours in the
flower have evolved in order to attract insects to pollinate
it is interesting; that means insects can see the colours.
That adds a question: does this aesthetic sense we have
also exist in lower forms of life? There are all kinds of
interesting questions that come from a knowledge of
science, which only adds to the excitement and mystery
and awe of a flower. It only adds. I don't understand
how it subtracts." Well, this could be a scoring point to
provoke your artist friends.

The Murmuring Brook

At some time or other in your life you
must have spent a sunny afternoon lying on grass,
listening to the murmur of a brook. It has a lyrical quality
that has evoked creative responses in many a poet and
musician. Do you know why the brook murmurs?

V Fly
One of the most beautiful sights is that of migrating birds flying across the evening sky in V formations. Why do migratory birds fly in V formations?

Light and Shade
In the shade of trees the ground is spotted with light—a common sight. Do you know why all spots are elliptical in shape, though their sizes differ appreciably?

Wet Bottomed
The ice at the bottom of an enormous glacier melts while the rest of the glacier remains frozen. Why?

Raman Confronts Rayleigh

In 1910, following a holiday at sea, Lord Rayleigh wrote: "The much admired dark blue of the deep sea has nothing to do with the colour of the sea, but is simply the blue of the sky seen by reflection". Eleven years later, during his voyage to England, the Indian Nobel laureate physicist Sir C V Raman, fascinated by the deep blue of the Mediterranean, started reflecting on the colour of the sea. He did a simple experiment on board to test Rayleigh's contention. Guess what did he do and the conclusions he drew?

Shades of Blue and Green

Seas do not have uniform colours. What factors determine their colours?

The Blue Dome of Air

The usual argument explaining why the sky looks blue is based on the fact that the intensity of light scattered by the particles in the atmosphere increases rapidly with decreasing wavelength (Rayleigh's Law of Scattering). Since blue light has a smaller wavelength than red, it is scattered more than red and hence the sky looks blue. But violet has an even smaller wavelength than blue. Then why doesn't the sky look violet?

Why Peaks Peak

Have you ever wondered whether there could be a mountain on our planet significantly higher than Mount Everest? The answer, surprisingly, is no! Why not?

Frozen Over

Lakes and rivers freeze in winter, yet aquatic life in them is not destroyed. How?

In Dribs and Drabs

Why does rain fall in drops and not all at once?

Footsteps on Sand
Why does the wet surface of a sandy beach dry up when we step on it?

Murmur in Sea-Shells
Is it really the sea you hear in a sea-shell?

Migratory Birds
Every year migratory birds fly across enormous distances to hospitable environments when their own gets hostile. How do these birds find their way every year like clockwork?

White Surf
"There are the rushing waves
mountains of molecules
each stupidly minding its own business
trillions apart
yet forming white surf in unison".

This is an excerpt from a poem composed by Richard Feynman, the celebrated physicist, for a public lecture he gave at the 1955 autumn meeting at the National Academy of Sciences, USA. Well, have you ever wondered what makes the surf that we see on breaking waves so bright and white?

Eelectricity
You must have heard of electric eels which can produce nearly 1 ampere currents at 600 volts or so. How do they do that?

A Flash of Lightning
What causes lightning?

The Chameleon Moon

The changing colour of the moon has evoked many a poet such as Shakespeare in *Romeo and Juliet*. By day the moon is striking pure white, while in the evening it becomes yellower, ultimately becoming pure yellow and then turning into yellow–white. Any idea why this variation occurs?

Brighter than the Sky

On a typically overcast winter day you must have observed that your snow-covered lawn appears to be much brighter than the sky, although the only source of illumination is the cloud light itself. How is this possible?

The Colour of Smoke

While walking along a busy canal, watch out for passing boats with oil or petrol engines that emit a fine smoke. When seen against a bright sky, it appears yellowish red. However, seen against a dark background, it looks blue. Why?

Dew Point

Why does dew form mostly on clear nights? Why do polished metal surfaces collect much less dew than materials such as glass in the same environment?

The Winter Veil

Winter is a season of smog in many places. While travelling across countryside by dusk we often see dense smoke hanging low over the tiled roofs. This is a familiar sight in the late autumn and winter but not in other seasons. Can you figure out why?

The Ghostly Moon

During a total lunar eclipse when the earth's shadow totally eclipses the moon, it is still visible and looks faintly reddish. Why?

Catch a Full Rainbow

On one occasion while travelling in a plane we happened to look down and saw a beautiful sight—a complete and circular rainbow with the plane's shadow (on the clouds below) at the centre! Why did we see a full circular rainbow?

The Moon and the River

One of our colleagues was recently flying on a moonlit night high above a river. He looked out through the window and noticed to his utter surprise that the moon's reflection on the river was so large that it did not fit into the width of the river! What puzzled him was that the width of the river appeared to have decreased with altitude as expected, but not the moon's reflection. What could that be due to?

Ignorance is Bliss

You must have seen birds happily sitting on dangerous high tension electrical lines. Why don't they get electrocuted?

Buzzing Bees

How do bees buzz?

The Elusive Cricket

Have you ever tried to listen to a cricket and locate it? The moment you think you hear the sound coming from a particular direction and turn your eyes towards it, it seems instantly to jump away to give you the slip. How do you account for this strange elusive character of a cricket?

Pondskater

Insects darting and skating along the surface of ponds are a pretty sight. How do they manage to do that without sinking?

Sap in the Cap

How does sap move up tall trees? As is well known, a vacuum pump cannot lift water columns beyond 33 feet because atmospheric pressure simply cannot support a taller column. Yet many trees are more than 33 feet tall. Some are two to three hundred feet tall. How are they able to pull water from the ground to their crowns?

Darkness at Noon

Breakers continually washing the shores are a beautiful sight. As the water rolls in and out, it leaves a mark on the beach. Wet sand looks distinctly darker than dry sand. Why?

The Shape of Ripples

When a stone is dropped into still water, it produces circular waves that spread outwards. What shape, do you think, will the waves take in the flowing water of a stream?

Twinkle, Twinkle, Little Star

"Twinkle, twinkle little star,
How I wonder what you are,
Up above the world so high,
Like a diamond in the sky".

Why do only stars twinkle but not planets?

The Blue Zenith

Have you ever noticed that the zenith (overhead sky) turns deep blue just after sunset? Any idea why?

Once in a Blue Moon

You must surely be familiar with the phrase "once in a blue moon". Have you ever seen a moon or a sun which is deep blue like the sky? Well, it is indeed an extraordinarily rare sight. To the best of our knowledge, a blue moon and a blue sun were first authentically reported way back in September 1950. Robert Wilson, an astronomer attached to the Royal Observatory in Britain saw a blue moon and also a blue sun in Edinburgh. He even made observations with a telescope and drew the strange inference that the blue of the sun and moon were related to forest fires in Canada. What could forest fires possibly have to do with a blue moon?

Halo Moon

Have you ever seen a halo around the moon? You must have. Do you know what causes the halo to appear?

Olbers' Paradox

When we look up to the sky at night, we find it pitch dark excepting for the stars. Can you guess what the darkness of the night sky is telling us about the universe we live in?

We are so used to the dark night sky that it is difficult to realize that this is indeed a profound puzzle. This is usually referred to as Olbers' Paradox after Heinrich Olbers, the German astronomer, who wrote an important paper discussing the puzzle in 1823.

The "paradox" rests upon the following assumptions : (a) The universe is infinitely extended in space ; (b) on the average there is a uniform distribution of stars and galaxies; (c) this distribution does not change with time; that is, the universe is static; (d) the universe is infinitely old; and (e) there is no matter in the intervening space to absorb and obscure the light from distant sources. From these assumptions, with a bit of mathematics, Olbers deduced that the sky should always be infinitely bright provided each source was a point source. If each source was an extended object, typically like the sun, and one takes into account the blocking of the light from more distant sources by nearer ones, the night sky brightness should at least match the surface brightness of the sun. Yet, the night is dark. Herein lies the "paradox" (scientists usually use the term "paradox" for "any plausible argument from plausible premises to an implausible conclusion"). Reality of the dark night sky is therefore telling us that some of the assumptions used in Olbers' reasoning must be wrong. Which ones do you think?

Give your Brains a Racking

❝A scientist has a lot of experience with ignorance and doubt and uncertainty. . . . We have found it of paramount importance that in order to progress we must recognize our ignorance and leave room for doubt. Scientific knowledge is a body of statements of varying degrees of certainty—some most unsure, some nearly sure, but none absolutely certain. . . . It is our responsibility as scientists, knowing the great progress which comes from a satisfactory philosophy of ignorance, to teach how doubt is not to be feared but welcomed and discussed. ❞

<small>RICHARD FEYNMAN</small>

**"Rouse up, Sirs!
Give your brains a racking
To find the remedy we're
lacking. "**

THE PIED PIPER OF HAMELIN

In this section we present a few intriguing riddles for you to mull over. We shall only mention some highlights of these problems; you have to figure out the complete explanations.

Richard Feynman (1918–1988) was born in New York City. He shared the 1965 Nobel Prize for physics for his work on the interaction of light with electrons. He was an outstanding teacher, managing to combine deep understanding of physics with great originality of presentation. His adventures outside physics are legendary. You can read about them in his book Surely You're Joking, Mr Feynman! Adventures of a Curious Character. *Photograph courtesy of AIP Neils Bohr Library.*

Feynman and the Wobbling Plate

In his famous memoir "Surely You're Joking, Mr. Feynman!" (W W Norton, 1985) Feynman recalls how one day he saw a guy in the Cornell University cafeteria fool around and throw a plate in the air. He noticed the red medallion of Cornell on the plate go around faster than the wobbling. This made him think and "play" with its physics. He writes : "The diagrams and the whole business that I got the Nobel prize for came from that fiddling around with the wobbling plate". He adds: "I discovered that when the angle is very slight, the medallion rotates twice as fast as the wobble rate". However, Benjamin Fong Chao in a letter to *Physics Today* (February 1989) has argued that the correct answer should be the other way round—a plate wobbles twice as fast as it spins when the wobble angle is small. Chao remarks: "Whether this error is a mere slip in memory, or, in keeping with the spirit of the author and the book, another practical joke meant for those who do physics without experimenting, we do not know and perhaps never will". Well, have a go at this problem, and if you are able to find a satisfactory answer which you can defend, please let us know.

Whispering Galleries

If you have visited St Paul's Cathedral in London you must have been struck by a large gallery in its dome which exhibits the special property of enabling faint sounds to be heard across large distances. Such structures are known as "whispering galleries". Scientists of the class of Lord Rayleigh and Sir C V Raman had spent considerable time in the 1920s trying to understand this phenomenon which entertains crowds of tourists but seldom provokes thorough scientific study.

A clue to its explanation is provided by the fact that you will hear your friend's whisper better the closer he/she is to the wall of the gallery. Also, a sound such as that of a handclap is heard over and again, the successive returns of the sound softening in its sharpness.

Lord Rayleigh's *The Theory of Sound* (Dover, New York, 1945 edition, pp 126–9) contained a concise discussion of this effect. But still further study continued. For example, Y Sato (*Nature*, Volume 189, p 475, 1961) felt it necessary to give a critical analysis of the explanation given by Rayleigh and Raman. If you can introduce innovative changes in the theory of this effect, you might well be able to design a modern marvel to excel Sir Christopher Wren's masterpiece!

Whistle a Melody

The physics of so common a phenomenon as whistling turns out to be quite intricate. Whistling is produced by what is called a hole-tone effect. When air with sufficient speed passes through a hole, vortices are formed and these ultimately produce the sound. However, the details of this process are not so clear.

The tea kettle whistle is another familiar example of the hole-tone. Such a whistle consists of two holes separated by a small cavity. When the stream of air from one end impinges on the other hole, vortices form which make the air enclosed by the second hole vibrate like the diaphragm of a loud speaker. If you are interested, you might like to read the review of the physics of the hole-tone effect, as far as it is understood, by R C Canaud, *Scientific American*, Volume 222, p 40, 1970. Have you ever tried whistling under water? Is it possible?

Feynman and the Tumbling Can

One day Richard Feynman came into the kitchen where his teacher John Wheeler's wife was cooking dinner. He took an unopened tin can and said to the children: "I can tell you whether what's inside is solid or liquid without opening it or looking at the label". "How?" asked the incredulous children. Feynman tossed the can up and watched it turn and wobble. "Liquid", he announced. He was indeed found to be right on opening the can. How did he figure it out? Clue: Feynman's "trick" worked because the liquid did not completely fill the can.

Hop Along

Have you ever skipped a stone on water? If so, you would have noticed that it bounces in a series of successively shorter leaps before stopping and sinking. E H Wright and K K Kriston discovered a queer phenomenon while skipping a stone on hard-packed, wet sand on a sea beach (*Scientific American*, Volume 219, p112, August 1968). The first bounce of the stone was short, the next a little longer, and then strangely, this short–long sequence repeated itself (periodic behaviour) until the stone came to rest. This behaviour has been found to occur with all stones of a regular shape.

It is interesting that the so-called "bouncing bomb" used during the Second World War for bombing dams utilised a similar effect. You may read about this in *The Royal Air Force in World War II*, edited by G Lyall (W Morrow and Co.,1968).

The peculiar character of the skipped stone's trajectory could be the result of a periodic interchange of translational and rotational kinetic energy of the stone. The fact that the coefficient of friction for sand is very high ought also to be an important factor, apart from the angles at which the stone approaches the sand and bounces from it. A detailed analysis of the physics of this phenomenon would be quite instructive.

Tippy Top

Photograph *courtesy of Neils Bohr Archive.*

Look at the photograph reproduced here. Can you recognise the two bewildered men? Well, they are Niels Bohr and Wolfgang Pauli, two of the outstanding physicists of this century. And the toy attracting their attention is the well-known "tippy-top". If you spin this top on its heavy spherical side, it quickly inverts itself and continues to spin steadily on its thin stem. But isn't that baffling? As H B G Casimir recounts in his article in *Niels Bohr—A Centenary Volume,* ed A P French and P J Kennedy, Harvard University Press (1985), Bohr used to often amuse himself with the physics of this toy. Casimir was told that the same thing also happens with a hard-boiled egg.

The steady spinning of a top is essentially the result of its inertia of rotational motion, referred to as the conservation of angular momentum. The flipping of the "tippy-top" does not violate this principle! Friction probably plays an important role here. However, why doesn't it keep flipping again and again? Why is the position with the heavier side on the top more stable during spinning? Think about it.

Of Floating Blades and Wooden Sticks
Take two razor blades and two wooden sticks. First, put the blades gently in a tub filled with water so that they do not sink. Now, slowly bring the blades towards each other by giving a slight push to one of them with your finger. You will find that when the blades are 3–4 mm apart, they will automatically get attracted towards each other and will remain stuck till you part them. The same phenomenon is observed, if instead of two blades, two wooden sticks are used. However, if you put one blade and one wooden stick in the tub, they repel each other if you try to bring them closer.

We urge you to repeat this experiment carefully with different materials and shapes and try to see if a general pattern can be established. Why do floating objects behave as they do?

The Mystery of the Floating Cork

Take a small piece of flat cork, wet it in water, and float it in a glass partially filled with water. You will find that it invariably drifts towards the walls. Keep pouring water gently. As soon as the glass is full to the brim, the cork automatically drifts towards the centre of the glass and remains there. Obviously, this is connected with the reversal of the curvature of the water surface. Why does this reversal take place? What precisely makes the cork drift towards the centre and stay there? What is particularly mysterious is that both when the glass is full to the brim and not, the cork tends to climb up the water surface! What is going on?

Red Star Over . . .

It is a remarkable fact that while a glowing object at a very high temperature (say, about 2000°C) looks white-hot on earth, a star of the same temperature appears to us reddish. Why?

A probable reason is that at the level of low intensities of star light received by us, the sensitivity of our eye is higher for red light than the other colours. Can you think of any other reason? A related phenomenon is that to our naked eyes most stars appear to have more-or-less the same colour, yet a coloured photograph taken with an optical telescope reveals a variety of colours.

Tap Dancing

Turn on the tap and close it slowly until you get a very thin but steady flow of water through the faucet. Place your finger in the stream and a standing wave-like pattern appears. Note that the periodicity in the pattern depends on the distance between your finger and the faucet. We do not know of any convincing explanation of this puzzling effect. Can you help?

A Vibrating Rainbow

In his book *The Nature of Light and Colour in the Open Air* (Dover, 1954) M Minnaert has mentioned a striking effect discovered by J W Laine—each time it thundered, the boundaries of the colours in a rainbow became obliterated ; "It was as if the whole rainbow vibrated". More studies have confirmed that this optical effect does not occur simultaneously with the lightning, but several seconds later, together with the sound of the thunder. The effect has been found to be correlated with an increase in the size of the water drops giving rise to the rainbow.

It has been suggested that the electric discharge during lightning could cause a change in the surface tension of the drops so that they coalesce more easily. This would, however, imply a strange coincidence between the time required for this change and the interval between the lightning and the hearing of the thunder. We leave it to you to judge.

The Blue Mountains

While travelling to hilly terrains, have you ever wondered why distant mountains often look blue?

One possible answer is that if the mountains are covered by green vegetation, they will absorb the redder part of the sunlight. Moreover, many hills have trees whose leaves emit volatile aromatic compounds. Consequently, they are enveloped by these vapours. The scattered light from this envelope is rich in blue light.

Our friend Unnikrishnan has pointed out that it is the distant hill in the shade of the clouds or in the morning/evening light (when direct sunlight is less) that appears more blue. Even distant vegetation in shade also looks blue. Unnikrishnan therefore makes the point that what matters is the relative magnitude of the intensities of the scattered light from the intervening space between the observer and the distant object and the reflected light from the object. Do you agree?

The Receding Blue

If you stand on a beach and look into the sea, you will observe that there is a sharp border near the horizon beyond which the sea looks distinctly more blue. If there is a cliff nearby and you start climbing it, you will find that this border appears to recede towards the horizon. Why?

A probable clue could be the fact that when we look at the sea from the top of a cliff, the intervening layers of air, illuminated by sunlight from the top or sideways, scatter predominantly blue light into our eyes. This is superposed on the blue background of the sea and could make the distant part of the sea appear more blue. But how this gives rise to the sharp boundary near the horizon which keeps shifting as we climb higher needs to be worked out.

Telltale Trails

You must have seen white trails on the sky left in the wake of a jet aircraft. The usual explanation one hears is as follows. Jet engines work by releasing high velocity exhaust gases. One of the combustion products of petrol and other aircraft fuels is water vapour. Since the atmospheric temperature at higher altitudes is well below 0°C, this exhaust vapour condenses into ice crystals which we see as jet trails.

While this explanation is not wrong, it is incomplete. If it were the whole truth, we would see these trails whenever the sky is clear, because the temperatures at heights where jet aircraft usually fly (around 30,000 feet) are well below 0°C. There must therefore be something else involved which prevents these trails forming in most cases. What could it be?

Larger Looms the Moon

You must have noticed that the moon appears larger near the horizon than when it is at the zenith. Have you ever thought about it? Well, it is certainly an optical illusion but what produces it remains a controversial issue.

L Kaufman and I Rock (*Science,* Volume 136, p 953, 1962; *Scientific American,* Volume 207, p 120, 1962) made a detailed study showing that the *apparent* enlargement at the horizon is 1.2 to 1.5 times the lunar size at the zenith. It has been observed that the effect persists in all atmospheric conditions. F Restle (*Science,* Volume 167, p 1092, 1970) has offered an amusing explanation based on an interplay between physics and psychology. If you are interested, you could take a look at it and try to form your own opinion.

Work Isn't Work

In physics, work done by a constant force is defined as the product of the force and the displacement in the direction of the force. When a man walks on a horizontal road at a constant speed, the pull of gravity is vertical and so the work done is zero. However, one does feel tired after a walk. Where is the catch?

The answer lies in the way we walk and the functioning of our muscles. In every step of a walk, the centre of gravity of the body is raised and lowered. If you hold a heavy book in your hand and raise it and lower it you will feel the strain because your hand muscles spend energy in this process. In the same way, the leg muscles exhaust energy with each footstep we take.

The difference between physiological work and "work" as defined in physics, however, is most clearly seen in the case of holding a heavy weight *stationary* on a stretched hand—you do feel tired though there is no displacement in the direction of the applied force by your hand. This is again because of the physiological strain suffered by the hand muscles. Surely you can think of many other examples. What would be the most general definition of physiological work? Any idea?

Hotter Freezes Faster

In cold countries it is observed that water left in the open during winter freezes faster if it is initially heated to a higher temperature. Isn't that counter-intuitive? It is interesting that even Francis Bacon had commented on this phenomenon. G S Kell made a systematic study and found that the effect is more pronounced if one uses wooden rather than metallic vessels without lids (*American Journal of Physics*, Volume 37, p 564, 1969). You may try the following experiment. Heat some water and pour it into a wooden or plastic vessel without a lid. Take the same amount of water at room temperature in a similar vessel, and put them both into the deep freeze of your refrigerator. You will find that the heated water freezes first.

One factor might appear to be the faster rate of evaporation of the heated water and the consequent greater loss of mass. If one starts with the same amount of water, by the time the temperature is equalized, the hotter water has lost more mass and so has a lower heat capacity. This means that its temperature changes more for a given loss (or gain) of heat. Subsequently, it cools down faster and can overtake the other in the freezing race. However, the loss of mass is perhaps not so significant. To test this, try the following variant. Take a larger amount of hot water and a smaller amount of cold water. Which one will freeze earlier?

Other factors which could effect the result are the size and nature of the containers, the temperature difference, convection rate and so on. It should be worthwhile to make an in-depth study of the relative importance of the various factors involved.

Glow by Night

Every once in a while, controversy arises about the claim that ice-fields on which the sun has shone for a long time glow by night. Snow is also sometimes said to glow when brought into a dark room from the sun. Could this be a genuine physical effect? Or is it an optical illusion? Obviously, a lot more work is needed before one can give a definitive answer.

Swimming in Circles

Richard Feynman was one day chatting with a group of swimmers. He heard them say that it helps to swim faster if the legs are shaved. Feynman got curious and wanted to verify whether this was indeed true. He suggested an ingenious test: a swimmer should swim in circles if one of his/her legs is shaven, provided shaving does indeed help to swim faster. Do you think that this test will work?

The Swing of Swings

Cricket fans will surely recall England's batting woes against Pakistan during the 1992 test series. A key factor was the devastating swing bowling by Wasim Akram and Waqar Younis who could produce "reverse swings" even at moderate speeds around 65–70 miles per hour (a "reverse swing" is a swerve in a direction opposite to what one expects from the bowling action). Many experts believed that appreciable "reverse swings" could never be achieved at such speeds. How did then Akram and Younis manage it?

The phenomenon of swing is indeed a complex affair involving a variety of factors like the bowler's grip, the speed and condition of the ball, atmospheric humidity and wind direction. The direction of swing, outswing or inswing, depends critically on how the shiny side of the ball is held by the bowler and his bowling action. A careful analysis of the physics of swings, including the "paradox" of the Akram–Younis performance, can be found in the article by William Brown and Rabi Mehta in *New Scientist*, 21 August 1993, pp 21–4. Research at Imperial College, London and University of Hertfordshire, Hatfield revealed that there is no magic about "reverse swings"; it can be achieved by any medium-fast swing bowler provided he allows the ball to scuff up on one side and then bowls with this side forward rather than the smooth side. Brown and Mehta explain the aerodynamics involved and conclude: "The key to conventional swing is to have one side as smooth as possible; the key to reverse swing is to have one side as rough as possible". It is also important to keep the angle between the seam and the line of flight as wide as possible. As the ball moves, a laminar boundary layer of air sticks to it because of viscous drag. As the ball's speed increases, it causes turbulence in this boundary layer, and this happens more easily on the rougher side than on the smoother side. Brown and Mehta argue that above a certain speed, this turbulence can generate a side force in the "wrong" direction, i.e., from the seam side to the smooth side. Maybe you would like to try to punch holes into their argument.

Meandering Rivers

Rivers do not normally flow straight, nor do they always bend in the same direction—they invariably meander back and forth. It is interesting that the celebrated physicist Max Born made the following comment on this: "The explanation of the phenomenon by means of the so-called "Coriolis force,"which the rotation of the earth exerts upon bodies which have a component of motion in a south–north or north–south direction, is trivial and universally known" (*The Born–Einstein Letters*, Walker, 1971, p 141).

Note Born's use of the word "trivial" which has a touch of irony because he implies rather erroneously that an object on the earth's surface must necessarily have a component of motion in the north–south direction in order to experience the Coriolis force. To recall the basic feature of this force, it arises in a rotating frame (such as the earth) and always acts in a direction perpendicular to both the velocity of the object and the angular velocity of rotation of the frame. It acts on all moving objects on the earth, *except* those which move parallel to the earth's axis of rotation, such as objects moving north–south or south–north at the equator. If the meandering of rivers is really due to the Coriolis force, rivers should always go round in circles like cyclones. But then why should a river turn one way and then the other?

Could a physicist of the rank of Born slip up on such an apparently simple matter? If it is not the Coriolis force, then what makes a river meander?

In a letter to *New Scientist* (2 May 1992) John Smith mentions that in the last century James Thomson (Lord Kelvin's brother) had pointed out that the meandering of rivers is essentially due to the "secondary flow" phenomenon (see answer to "Einstein in your Teacup", Chapter 1), although the situation here is a bit more complicated than in a tea cup because erosion at the bends must also play an important role.

Let us add yet another facet to this growing saga. You must have seen water gushing along a garden hose pipe. Why does the pipe wiggle like a meandering river? Is this similarity purely accidental?

Answers

❝The point is not to pocket the truth, but to chase it. ❞

ELIO VITTORINI

Richard Feynman and his father were once out for a walk in the woods when young Richard asked, "What is the name of that bird?" His father replied, "Well, I can tell it to you, but what is the use? We have one name, the Chinese another. But names are not essential in science. The essential things are how the bird uses its wings to fly, how he gets little ones, and how he came to be in the course of evolution. That is true science".

Kettle Croon

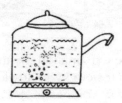

It is the bottom layer of water in the kettle that gets heated first. As the temperature rises, steam bubbles (not air bubbles) form at the bottom. Being lighter than water, they rise and come into contact with the cooler layers of water above, contract, and eventually collapse. It is the collapse of a myriad of steam bubbles that produces the hissing sound. The sound, therefore, increases as more and more steam bubbles form and collapse. Eventually, however, when the entire mass of water is heated to the boiling point, the steam bubbles do not collapse any more because they no longer encounter cooler layers of water. The hissing therefore ceases, and the whole mass of water in the kettle starts boiling.

Spoon in a Teacup

One puts in a metal spoon because metals are good conductors of heat. When hot tea is poured into a cup, the inner layers of the walls heat up and then gradually the outer layers. This uneven heating leads to uneven expansion and the cup cracks. Thick walls will therefore crack more easily than thin ones.

Einstein in your Teacup

The explanation mentioned by Schrödinger was discussed by Einstein in an article published in *Naturwissenschaften,* Volume 14, p 222, 1926 ; reprinted in *Einstein : A Centenary Volume,* edited by A P French (Harvard University Press, 1980): "The rotation of the liquid causes a centrifugal force to act on it. This in itself would give rise to a change in the flow of the liquid if the latter rotated like a solid body. But in the neighbourhood of the walls of the cup the liquid is restrained by friction, so that the angular velocity with which it rotates is less than in other places nearer the centre. In particular, the angular velocity of rotation, and therefore the centrifugal force, will be smaller near the bottom than higher up. The result of this will be a circular movement of the liquid of the type which goes on increasing until, under the influence of ground friction, it becomes stationary. The tea leaves are swept into the centre by circular movement and act as a proof of its existence".

Albert Einstein (1879–1955) in 1905 at age 26, the year in which he published his special theory of relativity and explanations of Brownian motion and the photoelectric effect, the last of which won him the 1921 Nobel Prize for physics. In 1916 Einstein completed his general theory of relativity of which Newton's classical theory of gravity is only a special case. Photograph *courtesy ETH Bibliothek.*

The crucial point here is that the differences in angular velocity within the rotating liquid (arising from viscosity) result in pressure differences (Bernoulli's principle). A pressure gradient develops along horizontal planes with the pressure increasing radially outward (the velocity is the least along the walls of the cup due to friction). In addition, a vertical pressure gradient also develops between the top and bottom layers because of friction with the bottom of the cup, slowing down the tea at the bottom compared to the top. These two pressure gradients set up what is known in hydrodynamics as the "secondary flow" within the liquid. While the tea is rotated, its surface becomes curved and the direction of the "secondary flow" is such that the tea leaves are driven away from the centre (if you observe carefully you will see that the tea leaves have a tendency to stay away from the centre). Once you withdraw the spoon and let the tea settle down, its surface begins to flatten out. The liquid mass readjusts itself and the pressure gradients decrease. This results in a reversal of the direction of the "secondary flow" bringing the tea leaves to the centre of the bottom (a detailed hydrodynamic explanation is rather complicated). It is indeed amazing that a simple everyday occurrence in a teacup contains such rich and complex physics.

A Hole in a Tea Pot

When a tea pot is filled with hot water and closed, the vapour gradually cools down and condenses, creating a partial vacuum inside. This makes the atmospheric pressure press down on the lid, making it difficult to open it. However, if there is a hole in the lid, air can pass through it neutralizing any difference between external and internal pressures.

The Teetotaller's Dilemma

The taste of tea does differ depending on whether milk is added to tea or vice versa. We understand from chemist friends that the main reason could be that chemical reactions in the two cases are different. In a tea cup, two types of chemical reactions ("denaturation" and "tanning") can occur which affect the protein part of the milk (known as casein) in different ways. Adding milk towards the end results in "denaturation" to a greater degree giving the tea what is known as a "boiled milk" taste. This can be avoided by putting in milk first. However, some of our physicist friends seem not to be satisfied with this explanation and they suggest that the origin of the difference in taste could be due to the fact that if milk is added last, it gets warmer than if the milk is added first. What do you think?

Fire Without Hazard

The fuel gas coming out of the holes in the burner is surrounded by oxygen in the air but cannot by itself catch fire. For the gas to burn we need to increase its temperature to its ignition point (360°C for the usual fuel gas n-butane) with the help of a gas lighter. Once a certain amount of gas reaches this temperature, the combustion process itself releases sufficient energy for the gas to continue to burn on its own in the presence of oxygen in the air. However, the flame cannot reach down to the cylinder because the fuel is stored in it as a liquid under high pressure. This pressure, being higher than the atmospheric pressure, prevents any oxygen from diffusing into the cylinder through the connecting pipe. Further, being a saturated vapour, its pressure remains the same almost until the entire fuel is used up. As a precautionary measure, a metal collar is usually provided at the mouth of the cylinder which can quickly conduct the heat away and prevent the fuel inside from being heated to its ignition point.

The Inner Core

Pure ice is transparent because of its homogeneity arising from the regular arrangement of ice crystals in it. The opaque inner core of an ice cube is due to inhomogeneities created by tiny air bubbles, usually less than half a millimetre in diameter, that get trapped between ice crystals during freezing. These air bubbles are formed because the solubility of the air dissolved in water decreases as the water is cooled. Since freezing starts at the walls of the tray, air bubbles get trapped in the central part which is the last to freeze. Light is totally internally reflected at the boundaries of the crystals surrounded by the air bubbles, resulting in a loss of transparency.

An Apple a Day

An apple contains tannic acid. When the cut surface of an apple comes in contact with air, the tannic acid reacts with atmospheric oxygen (oxidation) producing polyphenols which have a brownish colour. This can be avoided by sprinkling lemon juice on the surface. Lemon juice contains citric acid. A layer of citric acid on tannic acid prevents its oxidation.

Ovens with a Difference

Microwaves are electromagnetic waves having much longer wavelengths (of the order of centimetres) than visible light. A microwave oven works because organic food molecules are able to absorb microwaves at certain frequencies very well. In an ordinary cooker, the heat from the oven is mainly absorbed by the fat or oil used for cooking and the water in the foodstuff. In a microwave oven, heat is generated within the food itself by the microwaves agitating the food molecules directly. These intense molecular movements and collisions produce considerable heat within an extremely short time, enabling rapid cooking of food.

The container in which food is cooked in a microwave oven can be of plastic or glass but never metal. This is because microwaves cannot penetrate metallic containers. The glass or plastic container is transparent to microwaves and is not able to absorb microwaves at those frequencies to which the organic food molecules are highly absorptive. Hence such a container remains cold, though the food inside is hot.

However,one should be careful to ensure that microwaves do not leak out of the oven because they can have harmful effects on the human body.

Don't Lick an Ice Tray

There is always some moisture on your fingers. When you touch the frosted sides of the ice tray, this moisture freezes and the pressure of your fingers makes the frozen moisture stick to the ice crystals on the tray. If you try to lick the tray, your tongue will stick to it and a layer of the skin may be ripped off.

From Fermi to the Frying Pan

The answer lies in the simple fact that when food is fried, it is not the oil that boils but the water in the food, and, of course, the boiling point of water is lower than the melting point of tin!

Coiling Chocolate

The clues lie in high cohesivity and high viscosity of thick molten chocolate. High cohesivity makes it fall in a continuous stream without breaking into drops. The high viscosity prevents it from spreading too quickly on the plate after falling. This makes the initial bit of falling chocolate accumulate in a small heap which tends to keep its shape for a while. Subsequent streams of chocolate form distinct layers, one above the other, and these also retain their identities for a while before merging into a single heap. The layer on the top slips over the lower one and this makes it swirl. Incidentally, the same effect is also seen with shampoos.

Leaping Liquid

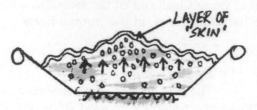

Milk consists of mostly water and some fats, proteins, lactose and minerals. Milk fat is a mixture of glycerides of fatty acids with a density less than that of the milk serum. The solid fat is dispersed in the serum in the form of small globules. These flat globules rise to the top and at a temperature around their melting point (about 50 °C) form a layer of "skin" on hot milk. The steam bubbles that form within the milk get trapped by this skin and accumulate under it. They grow and coalesce and build up a pressure that eventually raises the skin and makes some of the milk spill over. Stirring breaks the "skin", releases the pressure and prevents the spilling over.

Soup Swirl

Soup flow reversal is a simple illustration of what is technically known as "visco-elasticity". (An "ideal" fluid has no viscosity but all "real" fluids are viscous.) Once you stop swirling the soup, the layers in contact with the bowl come to rest because of friction.

However, other layers of the soup that are not in contact with the bowl continue to move. The stationary layers in contact with the bowl exert a visco-elastic restoring force on the moving layers which therefore slow down and eventually reverse their direction of motion. An oscillatory behaviour sets in, analogous to the oscillations of a spring which is stretched and released. These oscillations are eventually damped out by the soup's viscosity. If one is dealing with a fluid which is highly viscous, such as a paste, the oscillations can be damped to such an extent that only one reversal occurs.

Honey of a Problem

The neighbouring molecules of a liquid attract one another. Since a typical molecule well inside a liquid is surrounded on all sides by similar molecules, it is pulled equally in all directions. But the situation is different for molecules on the top. As a result, there is a net downward pull on these surface molecules which brings them closer to the molecules below them. But before long, a repulsive force begins to push them up and eventually a dynamic equilibrium is reached. This means that the molecules near a liquid surface possess additional potential energy (like a weight hanging from a stretched spring held vertically). This gives rise to what is known as "surface tension"—the liquid surface has stored in it a certain amount of potential energy per unit area. Since potential energy always tends to minimise itself, a liquid has an intrinsic tendency to minimise its surface area and shrink. This is why its surface acts like a stretched elastic membrane. (In the absence of gravity, a liquid always takes the form of a sphere because, for a given volume, a sphere has the minimum surface area.)

To facilitate discussions it is thus convenient to imagine an effective tangential force (the "surface tension force") along the surface.

Now to the "honey problem". As the weight of the accumulated honey exceeds the pull of surface tension, it lengthens and comes down. The slicing reduces the weight of the honey above the knife. If the slicing is not done too far down from the mouth of the pot, the surface tension is sufficient to overcome the pull of gravity and the honey is pulled back.

A similar phenomenon can be seen while watching drops of water falling from the mouth of a slightly open tap. The water drops lengthen as water accumulates at the mouth of the tap. As the drops get bigger, they become more and more elongated. They eventually break off from the tap and the remaining water shrinks back.

Have a Drink

When we drink, we first expand our chest with the help of our lungs. This expansion rarefies the air inside our mouth. Its pressure falls and the external atmospheric pressure forces the drink to enter into this region of lower pressure.

If you cover the mouth of a bottle containing a drink with your lips, you cannot suck in the drink because the pressure above the drink and inside your mouth is the same. You have to raise the bottle above your mouth and turn it upside down. Gravity then makes the drink flow down into your mouth.

Soap and Dirt

Dirt particles are of two types, oily and charged ones. Simply washing with water does not remove them because they tend to cling to our bodies and clothes. To make matters worse, oil does not mix with water. Soap molecules have the characteristic property (from their molecular structure) that they tend to get attached to oily and charged dirt particles. Subsequent washing with water then removes the dirt with the soap from the clothes.

Funny Funnel

As the liquid enters the bottle, it starts squeezing the air in it which cannot escape. This goes on until the air pressure in the bottle is high enough to hold up the weight of the liquid in the funnel. You must then lift the funnel a bit to let the compressed air escape. Then the liquid starts flowing down again.

Blow Out!

Newton's laws of cooling (not motion) are at work here. One of these laws says that the larger the difference between the temperature of a hot substance and its surroundings, the more rapidly it cools down. This is why, for example, our tea cools down quicker in winter. Also, when we blow over hot tea or milk to cool it quicker, we replace the hot air that accumulates over it by cooler air, and this helps more rapid cooling. There is another law of Newton which says that the larger the surface area of a hot substance, the more rapidly it cools. This is why pouring tea into a saucer helps to cool faster.

Both these laws operate in the case of the candle and cool the burning wax vapour below its ignition point (the temperature below which wax vapour does not burn). When we blow air at a candle flame, we (a) replace the hot air surrounding the flame by cooler air, and (b) distort the spherical shape of the burning wax vapour (not the flame which is not spherical) and increase its surface area. Simple geometry shows that for a given volume of a substance, the spherical shape has the least surface area. Any distortion from the spherical shape therefore increases the surface area of the substance and helps cooling.

Taking advantage of this latter feature, contrary to what you might expect, it is even possible to ignite coal or strip off paint by a jet of sufficiently hot air. Such contraptions (called "hot air strippers" or pokers) are now commercially available.

Iron it Softly

The starch in the cloth goes into solution when water is sprinkled on it. This helps to soften the cloth. A hot iron is useful because the heat helps to evaporate the water quickly, leaving a stiffened flat surface.

Fire! Fire!

There are two factors which make water a good fire extinguisher. First, water absorbs a large quantity of heat from the burning object (its specific heat is high). Secondly, the steam formed as the water boils in contact with the burning objects occupies a large volume and envelops the burning object, shutting off the oxygen supply. As you know, nothing can burn in the absence of oxygen.

Ice Fumes

When a large slab of ice is kept in the open, it gives out dense fumes. They are not fumes of any gas but simply water vapour that condenses in the cool air surrounding the ice. When the air surrounding the ice becomes very cold, some of the water vapour present in it condenses into tiny droplets of water. The condensed vapour looks like fumes when it moves up and down with the convection currents of air.

Coasting Along

The following points are relevant to this problem:

(a) The first point is that your hand holds the glass with the drink in it. The weight of the glass is therefore balanced and does not come into the picture.

(b) The thin layer of water between the bottom of the glass and the upper surface of the coaster removes all air from this region. Therefore the only pressure that acts downwards on the coaster comes from this layer of water. In order that the coaster gets lifted, the sum of the weights of this layer of water and the coaster must be less than the thrust of the atmosphere acting upwards on the bottom of the coaster. No wonder heavy coasters do not get lifted.

(c) The upper surface of the coaster must be fairly smooth so that no air bubbles can get trapped.

A Touch of Chill

In winter, our body temperature (which is usually around 37 °C) is considerably higher than the room temperature. So when we touch an object like a chair which is at room temperature, heat flows from our body to the object. Hence we feel it to be cold. Now, the faster an object carries heat away from our body the cooler it feels. Metals, being better conductors of heat than wood or plastic, carry away heat much faster, and so feel much cooler to our touch.

Tractors and Farmers

The answer lies in the difference between weight and pressure. Although the tractor is much heavier than the farmer, its weight is distributed over a much larger area of its bottom surface. Consequently, the load carried by each square centimetre of its bottom surface (the "pressure") is fairly low. On the other hand, the weight of the farmer is concentrated over a much smaller area of his feet, producing a much higher "pressure". An object penetrates deeper not because it is heavier but because it exerts a higher pressure (force per unit area) on its support.

Blinding Light

The human retina contains two types of light sensitive photo receptor cells called "rod cells" and "cone cells". Rod cells are adapted to sensing low light intensities but not colours and are useful in night vision. Cone cells are adapted to high light intensities and can sense colours. When light falls on the retina, the light energy is absorbed by a pigment (a protein called "rhodopsin", also known as "visual purple") in the photo receptor cells to yield a specific photochemical product which initiates the nerve impulses to our brain. The "visual cycle" is completed when the light sensitive component of the visual pigment is regenerated. We can adapt our eyes to a given intensity of the incident light by adjusting the size of the pupils and by controlling the eyelids so that there is a balance between the amount of pigment that is bleached by light and the amount that is regenerated. If the incident light intensity changes suddenly, this balance is disturbed, resulting in a temporary loss of vision until a new balance is achieved. If the incident light is too bright, we have to close our eyes altogether.

Rest in a Hammock

When you sit on a flat-topped stool, your weight presses down on a small area. A comfortable chair usually has a concave seat which helps to spread out your weight over a larger area. In other words, you exert less pressure per unit area. When we lie down on a soft bed, we make depressions that conform to the uneven shape of our body. Our weight is therefore more uniformly distributed, decreasing the pressure everywhere. This is why we feel so comfortable lying in a hammock or on a soft bed.

Long and Broken

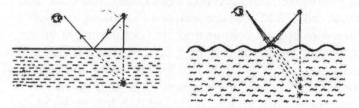

When the surface of a lake or pond is undisturbed, it behaves like a plane horizontal mirror. The law of reflection of light (angle of incidence equals angle of reflection) operates and only the light (from a point source on the opposite bank) reflected from a particular point of the surface can enter our eyes. This ensures that we see a clear image of the light source. However, when the surface becomes wavy due to the action of the wind, there are multiple points on it that are so inclined relative to us that they can all reflect the light into our eyes, and we see multiple images. As the waves move, these points also change and the images keep shifting.

Boot Polish

Polishing is such a mundane affair that we never bother to stop to think about it. Yet the answer is not so obvious. The surface of leather is full of hills and dales and fine hair. The dimensions of these irregularities are of the order of the wavelengths of light. Light can therefore "see" them and get scattered in all sorts of directions. This makes the surface look dull. The effect of the polish and the brushing is to even out the irregularities and make the surface "look" flat to light. The laws of reflection then make the surface look like a mirror.

Tear a Paper

Paper is made of cellulose fibres. When you tear a piece of paper, these fibres snap one after another and set off vibrations which produce sound waves in the surrounding air. When you tear it up quickly, you snap a larger number of these fibres in a given time and so increase the frequency of vibrations and hence the pitch of the sound.

Woof, it's Cold!

The temperature drops with height because of two factors:

(a) Although air absorbs all the dangerous rays from the sun (like ultraviolet, X-rays, etc), it does not absorb the sun's heat very much. It's the earth's surface which absorbs the sun's heat and warms the adjoining layers of air by convection.

(b) Normally, one would expect this heated air near the ground to expand and rise till it is on top of cooler layers. However, as it rises, it comes into contact with cooler and less dense layers and cools. Consequently, it cannot rise very far before meeting a layer that has the same temperature and pressure and is then trapped below the cooler layers above. Eventually, an equilibrium is achieved in which warmer layers of air remain trapped nearer the earth's surface. This vertical distribution, of course, has a daily and annual fluctuation.

The Foggy Mirror

The answer is quite simple. Put a bit of soap or detergent on the mirror (or the inside of the windscreen of your car during a heavy shower). A fresh slice of potato will also do. Now, what's the reason? It's all to do with surface tension and the angle of contact (the angle that a liquid drop makes with the surface on which it rests). No matter how clean you think your bathroom mirror is, it is in fact quite filthy. This is why the water that condenses on it cannot spread and wet it. Instead, it collects as small droplets. In other words, the contamination increases the angle of contact between the mirror and the water. No matter how hard you try, you are unlikely to be able to remove the filth completely, because even minute traces of it will affect the angle of contact. You can, however, use a thin coating of some liquid to reduce this angle. Detergents and the fresh juice from a potato do this trick. Without them, the minute droplets scatter light in all directions. This diffuse scattering causes the fogging.

Roll a Coin

When we place a coin vertically on its edge on a table, it is unstable because its base is small and a slight tilt makes the vertical line through its centre of gravity fall outside its base. It is similar to tight-rope walking. When we give the coin a push, we make it roll and acquire angular motion about an axis passing through its centre and perpendicular to its plane. Just as linear motion has inertia, angular motion too has inertia. If no external force acts on the coin tending to change its rotational state, it would continue to roll forever (the "conservation of angular momentum"). In practice, however, there is always some frictional force between the table and the coin, which slows it down and the coin eventually topples. But before toppling, it curves to the right or left. There is an instructive feature here. The turning of a coin to the right or left is, *in practice*, unpredictable, though its behaviour is, in principle, deterministic, i.e., governed by laws of motion which should determine its behaviour in a unique way. There are so many unknown and uncontrollable factors that can affect its motion (for example, slight defects along the edge of the coin, unevenness of the table, fluctuations in the breeze, sudden vibrations of the table and so on) that it is impossible to foresee them and take them fully into account. The moral is: determinism does not necessarily imply predictability *in practice*. This is a key feature of what is known as the phenomenon of "chaos".

Snoring Away

There is a soft flap at the back of our mouth. If a person sleeps on his/her back with their mouth open, this flap flutters back and forth because of the deep breathing. This produces the snore. Snoring can usually be stopped by closing the person's mouth and turning him/her over on one side.

Night Lends Clarity

Radio programmes are broadcast as medium or short waves. While medium waves travel parallel to the ground, short waves travel upwards through the atmosphere and are reflected back to the earth by the ionosphere.

The ionosphere consists of different ionised layers known as the D, E and F layers. The D and E layers exist only during the day when the sun shines. They usually disappear at night when the ions in them recombine to form neutral molecules. Those layers are not dense enough to reflect radio waves, but they can absorb a part of the incident energy. During the day, short waves which travel through these layers lose some energy before getting reflected by the F layer. This loss of energy results in a reduction of the signal strength. However, at night when the D and E layers are almost non-existent, signals can travel to far-off places without losing much energy.

Perfumes are airborne

Smell is a sensation created by the molecules of a volatile substance reaching the nerves in the nose. The volatile substance may be in the solid or liquid form. For example, perfumes are generally volatile oils, or aromatic substances dissolved in alcohol. The molecules of these substances pass easily into the air.

Even when the air in a room appears to be still, the molecules present in air constantly dart around, nudging the molecules of the perfume that come in their way. These perfume molecules get circulated throughout the room by a complicated diffusion process (*not* simple Brownian motion). Only a small number of them travelling across the room fairly quickly are sufficient to create the sense of smell. There are special cells in the nose (known as olfactory cells) to which these perfume molecules get attached and this in turn gives rise to an electric signal to a certain part of the brain which is responsible for the sensation of smell. The nature of smell depends on the characteristics of a perfume molecule and the way it gets attached to an olfactory cell.

The Yellow Fog

A foglight must penetrate as well as illuminate. Red light is most able to penetrate through air laden with fog, because it is scattered the least (among the colours making up white light) by the particles in the fog. This is why distant warning signals are invariably red. But red light has poor illumination characteristics. A driver needs not only to see warning signals but also his/her own way through. It turns out that due to evolutionary reasons, the human eye is most sensitive to yellow light which is most abundant in sun light. Not being far from red light in its penetrating power, yellow light provides the optimum combination of illumination and penetration for us.

The Painkiller Bottle

Certain fibres of the skin (known as pain fibres) are stimulated by the heat of the water bottle. This stimulus passes through the spinal cord to the affected muscle and also to its nearby blood vessels. These blood vessels are then dilated, and this helps to reduce the so-called "pain factor" in the muscle tissue. The "pain factor" gives rise to pain by producing toxic acids and by making the muscles contract and go into a spasm. Reducing the "pain factor" helps to relieve pain. A similar relief is achieved by a gentle massage.

Squeaky Chalks

While writing on a board a chalk stick is pressed against the board and made to move horizontally. The friction between the chalk and the board dislodges chalk particles which then stick to the board. If the friction is less than required, the chalk slips, touching the board several times in rapid succession. This gives rise to the squeaky noise. The force of friction between the chalk and the board depends mainly on the inclination of the chalk relative to the board and the surface areas in contact. Squealing occurs whenever the friction is small.

Raman's Billiard Ball Problem

The answer is surprising—it is the backward direction, i.e., the direction from which the striker ball drags the air around it. When it strikes the target ball, it stops momentarily. This makes the air trailing behind the striker ball get suddenly compressed and this produces a kind of shock wave with its intensity peaked backwards.

Play Cricket

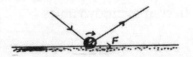

A bowler often delivers a ball with a spin in the forward direction. When a ball with its spin axis perpendicular to the vertical plane in which it moves hits the ground, there is friction between the surface of the ball and the ground. The friction opposes and reduces the spin. Some of the rotational kinetic energy thus lost goes into heating the surface of contact with the ground and a part gets converted into translational kinetic energy. This makes the ball move faster after pitching. Supposing the ball is spun backwards. What do you think will happen and why?

Top Spin

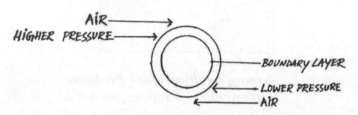

The essential point is as follows. When the ball spins, it carries with it a thin layer of air (boundary layer) which sticks to it. This causes a difference between the velocities of air flow at the top and bottom of the ball (see figure) and consequently a difference in pressure (Bernoulli's theorem). In order to make the ball dip, it is necessary to create a higher pressure and therefore a lower relative air velocity on the top. This can be ensured by making the ball spin forward relative to its direction of motion. This is an example of the Magnus effect in hydrodynamics.

Follow Shots

Although the translational kinetic energy of the cue ball is transferred to the other ball, it retains its rotational kinetic energy after a head-on collision. The cue ball therefore continues to rotate after the collision, slips for a brief while and eventually rolls forward because of the friction between it and the table. Energy of the ball is still conserved, except for the effects of friction.

Swimming Underwater

When we swim underwater, a layer of water covers the surface of our eyes. The refractive index of water is approximately the same as that of the substance of our eye lens. Hence no appreciable refraction can occur when light enters our eyes from the water. Consequently, no sharp images are formed on our retina and we cannot see properly. But if we wear goggles, then a layer of air is trapped between the water and our eyes. Air has an appreciably different refractive index from the material of our eye lens. This enables light rays to refract when entering our eyes and helps us see much better.

Ride Along

The answer to this problem involves "conservation of angular momentum". A cycle at rest is unstable because its base (the tyres) is narrow. A little tilt makes the vertical line through the centre of gravity fall outside its base and it topples. When you give it a rolling motion, its wheels acquire a rotatory motion. In the absence of friction, the inertia of angular motion (the angular velocity remains unchanged in the absence of any external force) would keep them moving. This gives them stability against falling. Once their speed decreases due to friction, they start wobbling and eventually fall. However, there are various aspects of bicycle-riding which continue to pose complicated problems in mechanics. For details you could take a look at a review of this problem by John Maddox in *Nature*, Volume 346, p 407, 1990.

Pole Vaulting

First, the polevaulter builds up kinetic energy by the long horizontal run-up. Then as he/she digs the pole into the wedge, a part of this horizontal momentum is converted into vertical momentum because the free end of the pole keeps moving due to inertia and describes an arc. The elasticity of the fibreglass pole enables it to bend backwards (by almost 90 degrees without breaking) under the weight of the vaulter. The potential energy stored in the pole during this deformation is then released as the pole unbends and this catapults the vaulter over the bar. Of course, the skill of the polevaulter plays an important role in determining the height he/she clears.

Sleek and Swift

Extensive and systematic research in modern times using advanced technology is trying to provide the right kinds of scientific input for creating new world records. For instance, wind tunnel experiments have revealed that wind resistance is a major hindrance, particularly to cyclists and sprinters. For example, in the case of a cyclist riding at 50km/hour on a smooth road, wind resistance accounts for almost 80 per cent of the drag. Such studies have shown that tight clothing and short hair help significantly to reduce air drag. Putting on a cap is an added help. It is reduced air drag that also makes a dimpled golf ball travel farther than a ball with a smooth surface.

The higher the speed, the more important is the air drag. Tight clothing therefore benefits the sprinter and the cyclist more than, say, marathon runners. Experiments have also shown that tight wool jerseys are much more effective in overcoming wind resistance than tight cotton jerseys.

Cyclopean Vision

The crucial point is that while the combination of two eyes enables us to judge depth or distance, single-eyed vision gives us better sense of alignment. Suppose one is aiming at a target with an arrow. With both eyes open, one cannot keep the arrow aimed at a target because the straight lines from our eyes to the target through the tip of the arrow make a small angle with each other and intersect at the tip. So the two eyes give us slightly different alignments. This is avoided by keeping one eye closed. However, the price we have to pay is the loss of the sense of distance or depth. It is therefore remarkable that Mansur Ali Khan Pataudi (who played for Sussex County and later captained the Indian cricket team) was a good batsman as well as a brilliant fielder in spite of having one-eyed vision.

To digress a bit, another curious thing is that a painting or photograph looks better with one eye. The effect is enhanced if we look through a tube. The reason is simple. When we look at a picture with both eyes from a moderate distance, we recognise it as a flat surface. But when we look at it with only one eye, our minds are free to take up the suggestions made by light and shade in the picture. So, after we have gazed at it with one eye for a while, it begins to acquire a three dimensional character.

Grand Jete

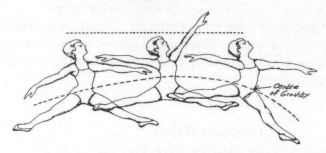

First of all, the dancer jumps in such a manner (at 45° to the ground) as to spend the longest time near a peak. During this period the dancer manipulates his/her head and the limbs in such a way that although the centre of gravity of the body moves along a parabolic path, the head maintains almost a straight line path (see the figure). Since we are likely to follow the dancer's head, an illusion is created of gliding. The legs are also raised and they execute a split at the peak, adding to the illusion of a glide.

Soaring High

When you stand, you exert a force equal to your weight on the ground. In turn, the ground exerts an equal and opposite force on you which is why you do not sink. This is an example of Newton's third law of motion which says that to every action there is an opposite and equal reaction.

When you start to walk, you exert a force on the ground that is inclined to the vertical. You could then regard the ground reaction on you as a combined effect of two forces, one vertical and one horizontal. When you start to run, faster and faster, you keep increasing the force with which you push the ground. As long as you do not slip, the horizontal component of the ground reaction remains balanced by friction and the *vertical* ground reaction on you keeps increasing. This is why a run-up before a high jump helps to provide a vertical push upwards.

The Juggler's Trick

The trick lies in giving the hat a spin about its axis of symmetry. If one spins a top, it stays stable about its vertical axis of rotation. If you try to tilt it, it would come back to its stable upright position. Rotating bodies invariably seek to maintain the direction about which they rotate. This is the law of inertia for rotational motion—once spun, an object keeps rotating with constant angular velocity about a fixed axis unless acted on by an external force. This is also known as the gyroscopic principle (gyroscopes are used in navigational devices). Another common application of this principle is in the design of the barrel of a gun. The barrel of a gun is rifled, i.e., it has spiral grooves cut inside which make a bullet leave the gun spinning about a well-defined axis. This gives it an additional directional stability. It is this same principle that makes a spinning hat come down the right way every time, enabling the juggler to catch it on his/her head.

Feat of Flying

The basic principle underlying lift is "Bernoulli's principle". Stated in simple terms, it implies that along any flow line in a fluid the transverse pressure will be lower if the flow velocity is higher. An aircraft wing is designed to have a convex top surface and an almost flat under-surface. This makes the streams of air above the wing move faster than those below it. This is because, being fairly incompressible, the same mass of air has to travel a greater distance in the same time to conserve mass and energy. Bernoulli's principle then tells us that there is higher pressure below the wing than above it, providing the required lift.

In practice, it is necessary to take into account the adhesion of air to the wings and viscous forces. The full theory of aerodynamic lift is rather complicated, but the net effect is to introduce a turbulent motion of airflow around the wing. This circulation turns out to be such that above the wing the circulation speed is added to the flow velocity of the air while underneath the wing it is subtracted, helping the lift further.

An effective flow of this air current can be generated by facing the wind, or by rushing through air, or by using a combination of both, that is, by running into the wind.

Of Birds and Aeroplanes
The key difference is that birds use their wings for both lift and propulsion, whereas aircraft use their wings for lift and their engines for propulsion. How an aircraft generates its lift has been explained earlier. Birds' wings are the last words in aerofoil design, and aerodesigners have not yet been able to come up with wings that are both flexible and strong to achieve lift as well as propulsion. A bird also requires a strong heart to power long flights. The powerful heart is kept warm by feathers, one of the best natural insulators one can think of. Thus, equipped with a superb engine, the world's best wings and springy toes for a takeoff, a bird can soar into a graceful flight at a moment's notice.

Smoky Swirls
The hot gases (smoke) from the burning cigarette first rise slowly and have a smooth laminar flow. They then accelerate because of the buoyant force exerted on them by the cooler surrounding air. After a few centimetres the velocity is high enough for turbulence to set in, and eddies or "vortex" air currents form.

The Fluttering Flag

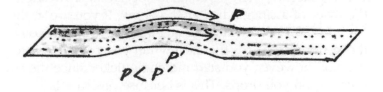

$$P < P'$$

Imagine the flag perfectly flat and fully
spread out in a strong wind. Suppose a small disturbance
develops in one part of the flag that causes a ripple in it.
The air stream flowing across the flag must speed up as it
crosses over the ripple. The faster moving air has less
pressure *P* (Bernoulli's principle) and hence there is a
difference in air pressure on the two sides of the flag
near the ripple. This happens randomly all over the flag.
It is these pressure differences that cause the flag to
flutter in the wind.

Pour a Liquid

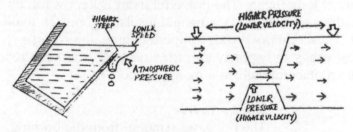

It is atmospheric pressure in conjunction
with Bernoulli's principle that makes a liquid stick and
run down the side of its container. The Bernoulli
principle is a very general principle of fluid flow and has
numerous applications. This principle is really a
consequence of the fact that energy can neither be
created nor destroyed (the fundamental principle of the
conservation of energy). Imagine spectators crowding in
the foyer of a cinema hall after a show. If you are in the
foyer you feel everyone pressing against you, and you
move slowly towards the exit. When you come close to
the exit, however, you start moving quicker since the
pressure on you drops. This is because pressure is
generated by the sideways pushes of people around you.

When everyone moves forward, this pressure drops. The same is true of the molecules of a liquid. When they move slowly, they jostle and collide against one another and the walls of the container, creating a pressure. When they approach a narrower section of the tube, they move forward faster because, being incompressible, the same amount of fluid has to pass through the narrower sections in the same time. The pressure consequently drops.

When you pour a liquid out of a thin walled container, the bottom layer of the stream in contact with the edge of the container turns round much faster than the top layer. According to the Bernoulli principle, therefore, there is a pressure drop across the width of the stream, and the atmosphere presses the stream against the container side.

If you pour out the liquid fast by turning the can quickly, the stream acquires an overall velocity, there is no longer any appreciable pressure drop across its width, and the liquid does not stick.

In order to prevent this sticking of liquids, juice bottles usually have a thick round skirt at the mouth which eliminates the difference between the curvatures of the bottom and top layers of the flowing liquid; there is therefore no pressure drop across it. Milk and tea pots have a spout or lip which makes the stream run down it and drop straight into the cup without sticking.

The Tapering Stream
There is no force that squeezes the water stream and makes it narrower as it falls. It is the conservation of mass at work. Since water is

incompressible, the same mass or equivalently the same volume of water must pass through every cross-section of the stream per second. Since water speeds up as it falls, more water would pass through successive cross-sections per second unless the cross-sectional area decreases. It is gravity pulling on the water that is ultimately responsible for the narrowing of the stream.

The Expanding Smoke Rings

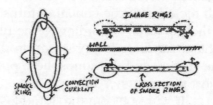

   Let us first see why smoke rings are stable far away from walls. The hot smoke ring (in a vertical plane) sets into motion convection currents in the surrounding air which thread the ring as shown in the figure. Since there is no preferred direction in the space surrounding the smoke ring, convection currents flow symmetrically all around it. The ring, therefore, experiences equal pushes and pulls from every direction. The net effect is nil, and the ring is stable. However, as it approaches a wall, the convection currents strike the wall. Since the layer of air in contact with the wall is nearly at rest (viscosity), the presence of the wall affects the convection currents which can no longer flow symmetrically around the ring. The proximity of the wall spoils the isotropy of the space surrounding the ring. The components of motion perpendicular to the wall get cancelled, while those parallel to it get reinforced.

Consequently, the ring expands. The delicate interplay between symmetry and dynamics in such a common phenomenon is indeed fascinating.

The Puzzling Balloons

There are two factors involved in the explanation of this problem: (a) Archimedes' principle, and (b) "pseudo-forces". A helium-filled balloon experiences an upward buoyant force equal to the weight of the air it displaces (Archimedes' principle). Since helium is less dense than air, this buoyant force is greater than the balloon's weight. It therefore floats upwards against gravity. Now, when a car accelerates, say in the forward direction, a backward force is generated on massive bodies inside it because of their inertia of rest. When a car brakes, the inertia of motion produces a forward force inside it. Such forces which occur in the accelerated frames of reference (such as an accelerated car) are called "pseudo-forces" to distinguish them from "impressed forces". Unlike "impressed forces", "pseudo-forces" do not arise from the action of other physical bodies. They act on objects in the accelerated frame, proportional to their mass and acceleration of the frame. Though the term "pseudo" is conventionally used here, there is nothing "unreal" about these forces.

The horizontal forward or backward pseudo-force inside the car generates the equivalent of a horizontal gravitational field (this "equivalence" is one of the most profound principles of physics, discovered by Einstein and popularized with the help of his famous "thought experiment" of a freely falling lift inside which gravity disappears). The helium-filled balloons being lighter than air will move against this "pseudo" gravitational field (Archimedes' principle).

An Anti-gravity Effect

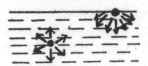 

The source of energy in capillarity is "surface tension". The rise of water in a capillary glass tube results from an intricate interplay between surface tension (intermolecular forces within water) and adhesion (intramolecular forces between water and glass).

Let us think of a liquid stored in a container. If the liquid wets the container (like water in a glass), its surface is found to curve upward near the edges and make a definite "angle of contact" as a result of an equilibrium between adhesion and cohesion. On the other hand, if the liquid does not wet the container (like mercury in glass), its surface curves downwards, again at a definite angle of contact (the shape of the curved surface of a liquid in equilibrium is called a meniscus).

When a capillary glass tube is inserted into water, its bore is so narrow that the angle of contact between water and glass cannot be stable (the formation of the meniscus is impeded) because the equilibrium between adhesion and cohesion is not maintained. The surface tension dominates over adhesion and pushes the water up the tube until gravity balances it.

Through the Palm, Strangely!

When we look at an object, both our eyes get focused on it automatically even when we keep one of them closed. This is called "sympathetic focusing" or "adaptation" of our eyes. In the experiment concerned, your left eye is focused on a distant object. In sympathy, your right eye also gets focused on it, although it is closed. When you bring your right palm in front of it and open your right eye, your palm appears blurred or de-focused. In other words, your left eye sees the distant object clearly through the tube while your right eye does not see the palm clearly. This gives you the impression that you are seeing the distant object through a hole in your right palm. In order to verify this, do the experiment again and try deliberately to look at your right palm. The moment you concentrate on it, the palm will come clearly into view and the distant object and the hole in the palm will disappear.

It Does Not Pour Out

There is always a tendency for the water to flow out of the glass through the space between the rim of the glass and the stiff card. To observe this, use a thin piece of metal in place of the card and press it against the inverted glass. You will see a thin layer of water bulging and skirting the glass rim. The moment you release the pressure on the plate, you will notice that this water will disappear into the glass and the plate drops a little. This is enough to make the air above the water inside the glass expand and make its pressure drop sufficiently so that the atmospheric pressure is able to hold the card with the water above it. To test that the air above the water does indeed exert less pressure than the atmosphere, you can make a hole in the bottom of the glass and fix a glass tube through it so that one end of it is in the space above the water. Do the experiment with the other end of the tube closed with a finger. As soon as you remove the finger, the card will drop.

Blow Hot, Blow Cold

When we blow air with our mouth wide open, the air in our lungs comes out without any noticeable expansion. This air has the temperature of our blood which is usually warmer than the ambient temperature. This is why our hands feel it as warm air. When we blow this air with our lips pursed (with a narrow opening), it suddenly expands. In expanding, it has to do work against the intermolecular attractive forces. In other words, it has to spend its own energy. Since the expansion is rapid, it has no time to absorb any energy from the environment. This causes a drop in its temperature. This is an example of the Joule–Thomson effect.

Through a Glass Darkly

Let us consider a diffusing screen placed between us and a point source. One should imagine a diffusing screen to be simply a piece of smooth glass covered with particles (larger than the wavelength of light) which scatter light in all directions but with the intensity peaked in the forward direction. As a result, we see the point source surrounded by a diffuse halo of light whose intensity falls radially from the centre. However, the size of the halo increases as the source is moved away from the screen.

Now suppose that a second point source is placed close to the first one. Then the presence of the screen will make the two sources appear less distinct if their haloes overlap. Since the haloes grow larger in size as the sources move away, the screen blurs their details more, the farther they are from the screen.

Incomprehensible Whispers

Whispers are characterized by sound waves of wavelengths much shorter than the wavelengths of the normal voice. This makes their ability to bend or "diffract" around our heads much less than that of the normal voice. When your friend is turned away, the whisper can be heard only through bending of the sound waves around his/her head (in the absence of reflection of the sound from any nearby object). Even a loud whisper uttered in this condition could therefore remain incomprehensible.

Falling Cats

All bodies through the earth's atmosphere accelerate up to a terminal speed determined by air friction (proportional to their surface area) and weight. Thereafter the terminal speed remains constant as long as the weight and surface area do not change. The weight, of course, cannot change, but one can change the surface area exposed to the air. Free-falling parachutists spread their arms and legs apart to increase this effective area of contact with air in order to brake their falling speed.

Falling cats behave much like parachutists. After reaching the terminal speed they spread their limbs horizontally like a flying squirrel. This increases the air resistance, decreases their falling speed, and spreads the impact on reaching the ground over their whole body. The risk factor is further minimised because of some additional features unique to cats. These include their flexed limbs which spread the shock of impact through their flexible joints and muscles as well as a superb in-built gyroscope located in their inner ear. If a cat begins to fall upside down, it is able quickly to twist in midair and reorient itself with all four legs pointing downward by the time it has fallen just about a metre. In contrast, human beings possess a much less efficient gyroscope and tend to tumble uncontrollably as they fall.

Ice in a Scarf

A woollen scarf is not able to give anybody any warmth in the way a glowing fireplace does. It merely prevents our body from losing its own warmth. That is why a warm blooded animal whose body is a source of heat feels warmer in a coat of fur than without one. The ice inside a warm cloth takes longer to melt because the warm cloth is a bad conductor of heat—it inhibits the flow of heat from the surroundings to the ice.

Dropping a Bottle

Since it is safer to jump off a moving bus or train facing the direction of motion, you might think that the bottle should be thrown forwards. You are wrong. It should be thrown backwards, because its velocity of projection would then be opposite to its inertial velocity (the velocity of the bus or train) with the result that it will strike the ground with a smaller impact. If you throw it forward, its velocity of projection will add up with its inertial velocity and it will strike the ground harder.

Why then is it safer for us to jump from a moving bus and run forward? The answer is that we then avoid falling flat on the ground and injuring ourselves.

The Burning Flame

The initial deflection backwards is due to the inertia of rest. However, contrary to expectations, the protected flame, when carried, will move forwards, not backwards. This is essentially because the hotter vapour of the flame is lighter than the surrounding trapped air. When a force is applied to a body to move it, the acceleration is faster, the smaller is the mass (Newton's second law of motion). Being lighter, the flame moves faster that the surrounding trapped air and is therefore seen to be deflected forwards.

Taper Caper

Let us first examine what happens when a candle burns. First, the wick catches fire and then the wax at its base melts and rises to the top of the wick by capillary action. It is then vaporized and the vapour catches fire. All this takes time. When a candle is blown out, the region surrounding the wick remains hot for a while and a bit of molten wax continues to trickle up and vaporize. If the burning splint is brought near the hot vapour, it catches fire instantly.

Wet a Brush

The reason is simply because a surface of water present on the brush hairs makes the hairs cling together through the action of surface tension (the tendency of a water surface to shrink in order to minimise the potential energy stored in the surface).

Tyger! Tyger! Burning Bright

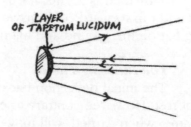

Unlike human eyes, a cat's eyes contain crystals of Tapetum Lucidum which reflect light. There is a layer of this substance behind the retina of a cat's eyes. This reflects back the incident light so that it passes twice through the retina, thus enabling the cat to see better in light that is too dim for us. Moreover, cats have many more rod cells than cone cells in their retina. Rod cells respond to brightness whereas cone cells recognize colours. This is why cats are practically colour blind but this is made up by their ability to see better than us in the dark.

Hum with your TV

When a person hums a particular pitch or frequency, his/her eyeballs start vibrating with the same frequency. How this occurs was first analysed in detail by W A H Rushton in an article published in *Nature*, Volume 216, pp 1173-5. He gave a physiological explanation of how humming affects the brain, and also suggested several experiments to demonstrate the effect.

Television pictures are formed by the recurrent line-by-line horizontal scanning by an electron beam which excites the screen. The frequency with which the electron beam sweeps the screen from top to bottom is so high that the recurring images appear continuous to our eyes. If a viewer hums with the same frequency, his eyes start opening and closing with that frequency, producing a stroboscopic image of the screen on his retina. In other words, the image freezes on the retina. If he/she hums with a frequency too high or too low, the image will appear to move upwards or downwards. Obviously, the effect is only visible to the viewer who hums.

Play on a Ship

Neither has any advantage if the ship moves steadily in a straight line. You might think that the person standing nearer the bows recedes from the ball after it is thrown and the other person moves forward to receive it. A little reflection will show that this is not true. The ball as well as the two friends are carried by the ship and therefore have the same speed as the ship (inertial speed). Therefore the ship's motion (as long as it is steady and in a straight line) cannot give any one of them an advantage over the other.

In fact, to all passengers on board such a ship, everything would proceed as if the ship were at rest and the water and the shore were moving in the opposite direction. There is no physical way of distinguishing uniform velocity from rest. Uniform velocity is purely relative. This is known as the "principle of relativity".

No Spilling Over

The correct answer is surprisingly very simple. A floating ice cube displaces water whose weight equals the weight of the ice cube. When the ice cube melts, the water formed has the same mass as that of the cube (principle of conservation of mass) and hence the same volume as the water displaced by the floating cube. The total volume of water in the glass therefore remains unchanged after melting.

To Catch a Card

The fact is, when you do the dropping and the catching yourself, both your hands get signals from the brain simultaneously, one saying "Let go !" and the other, "Catch !". Now, when someone else drops the card and you have to catch it, there is a time lag between your seeing the card released and your hand responding to a command from the brain to catch the card. This is why the card goes through every time.

Of course, if the card is long enough, you can always catch it. This tells us something about human reaction time. To find out your reaction time, experiment with cards of different length. Measure the length of the shortest card you can catch and use the value of the acceleration due to gravity to calculate the reaction time which should be of the order of some fraction of a second.

The Eclipse of Superstition

There is nothing special about sunrays during a solar eclipse. It is always harmful to stare at the sun with naked eyes, because our eyes will focus energy of the rays on to a point on the retina and burn the cells, causing serious damage. During a total solar eclipse when the ambient light is very weak, our eyes get adjusted to the low light intensity by opening up the apertures. In these conditions if one keeps looking at the sun, then as it emerges suddenly at the end of the eclipse, our reflex action in closing down the eye apertures is slow in comparison to this sudden change. Our eyes therefore let in more light energy than is good for them. This is why it is advisable to use a dark filter when watching a solar eclipse.

The Invisible Silver Thread

The mercury in a thermometer forms an extremely thin thread in a capillary. A beam of light falling on it is reflected almost entirely in one specific direction. If one's eyes happen to catch the reflected beam, the mercury thread looks silvery white. From every other direction the thread appears black or dark grey.

Which is Heavier?

The two glasses will weigh the same. This is because the floating piece of wood displaces exactly its own weight of water, and so, although the glass with the piece of wood has less water in it than the other glass, the weight of the piece of wood exactly balances this loss. Archimedes' principle again!

Tearing Wet Paper

It is the cohesive force between the cellulose fibres (of which paper is made) that must be overcome in tearing paper. In the presence of water, this cohesive force which is of electrostatic origin is weakened, in much the same way as solubles like salt (e.g., sodium chloride) dissolve in water because of the weakening of the electrostatic attraction between the positively and negatively charged ions. In the case of paper, the effect is easily perceptible because water wets paper and water molecules can flow into the spaces between the fibres, weakening the cohesive force between them.

The Jumping Draught

You must have noticed that when a moving ball is made to hit an identical ball that is stationary, the moving ball stops and the target ball rolls forward with its velocity. This is an example of an "impact". In this case the "impact" occurs between two elastic bodies. An impact lasts a split second. During this short time, however, a whole process occurs. First, the two bodies compress each other at the point of contact. Internal restoring forces are generated by this compression. When the compression reaches its maximum, these internal forces begin to push the bodies out in opposite directions and restore their shape. The moving ball is stopped by these restoring forces and its velocity is transferred to the target ball. We may say that the "impact" is, as it were, transferred from the first to the second ball. This is an example of two fundamental laws of mechanics—the conservation of energy and the conservation of momentum. Exactly the same thing happens with the draughts or coins. The "impact" is transferred from the first draught through the intermediate ones to the last one which has no other draught to transfer—so it moves away.

Weigh a Stone in Water

The balance is maintained. This is because, although the stone should weigh less in water than in air because of the greater upthrust the water exerts on it, it will also displace its own volume of water

whose level will rise. The water will then exert an
additional force on the bottom of the glass exactly equal
to the weight lost by the stone—again an illustration of
the Archimedes' principle.

Comb your Hair

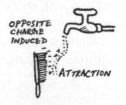

When you comb your hair or rub the
comb with a piece of flannel, the comb is weakly
electrified. The proximity of the comb induces an
opposite electric charge on water molecules. The comb
and the water therefore exert an electrical force on each
other. Since you hold the comb steady, it is the water that
gets deflected. The trickle becomes a steady stream
because of a change in the surface tension of water as a
result of electrification.

Weigh Yourself
When you bend forward, the muscles that
help you do that pull up the lower half of your body.
This is why your body exerts a lower pressure on the
weighing machine. When you lift up an arm, the muscles
used to do this push down on your shoulder and
increase the pressure on the weighing machine. Of
course, your mass does not change at all. It is Newton's
third law of motion that operates. The sudden motion (or
strictly speaking, the momentum) of your hand upward
must be balanced by an opposite movement downward.

Darting Pepper

The answer lies in the surface tension of water. You will recall that the surface of water behaves like a stretched rubber membrane, this helps pondskaters to move about on the surface of ponds without sinking. Well, soap lowers the surface tension of water. So, when you put a little detergent soap on the water surface, its surface tension is lowered locally. This is like making a hole on the surface of a stretched membrane—the punctured membrane shrinks, carrying the pepper with it.

The Puzzling Rubber Band

The quick stretching of a rubber band is an "adiabatic" process—a process in which no exchange of heat with the surrounding can occur. The work done by us in stretching the rubber band, therefore, goes entirely to increase its internal energy. This raises the temperature of the rubber band. On the other hand, when a gas expands rapidly, the gas itself has to do work against attractive intermolecular forces, and it draws the required energy from its own store of internal energy. Consequently, the gas cools down.

This simple phenomenon of the warming of a rubber band when suddenly stretched has a fascinating, albeit a bit more technical, facet. A rubber band is a disordered tangle of long chains of molecules. The stretching produces a more ordered arrangement of these molecular chains. In statistical mechanics, "entropy" is usually taken to be a measure of the disorder of a system. So one would expect the entropy of the band to decrease when stretched. But it follows from thermodynamics that entropy cannot change in an "adiabatic" process. The resolution of this paradox lies in recognising that the usual connection between entropy and disorder is not valid in non-equilibrium situations (situations far from equilibrium, as in a stretched rubber band). This notion is one of the key take-off points for what is known as "non-equilibrium thermodynamics", a subject that is only just beginning to be properly understood. Ilya Prigogine was awarded a Nobel prize for his pioneering contributions in founding this subject.

Not with a Bang but a Whimper

When a gun is fired, the explosive in the cartridge burns out rapidly, producing hot gases. These high pressure gases expand and eject the bullet and flow out into the air outside with high velocity. This produces a shock wave resulting in a bang. To cut down this noise the gas velocity has to be reduced. This is achieved by a silencer. It is a tube with a number of thin metal plates fitted coaxially in a row along the tube. Each plate has a hole at its centre to allow the bullet to pass through. This reduces the velocity of the gases following the bullet considerably.

Fahrenheit 451

451°F (about 232°C) is the ignition temperature of paper. This is the temperature to which paper must be heated to make it burn. Since this is much higher than the boiling point (212°F) of water under normal atmospheric pressure and the specific heat of water is very high, one can easily make water boil in an uncovered pot made of thin but stiff paper without burning it (a thin paper helps to conduct the heat through it quickly so that the paper does not get much heated).

Wait Until Dark

The night is never totally dark, so one way of "seeing" at night is with the help of a device which amplifies what little light there is. Such a device is called an image intensifier. A television camera fitted with an image intensifier can produce good quality pictures at night.

An image intensifier makes use of a thin layer of a certain type of "photoemissive" material which emits electrons depending on the intensity distribution of light falling on it. When these electrons are made to fall on a phosphor screen, many more photons are produced than those falling on the photoemissive layer. In this way a millionfold amplification can be obtained even by a pocket sized night vision binocular, widely used in the military.

An Oscar-Winning Problem

If you dip one end of a handkerchief in water, a large part of it gradually gets wet. This is because of capillary action. A handkerchief or any piece of cloth consists of a whole lot of capillary tubes with a fine bore. Water rises through these capillary tubes due to the action of the surface tension (a blotting paper also works on the same principle) and eventually moistens a large part of the cloth. Exactly the same thing happens when a crumpled cloth is thrown onto water. As it starts soaking water, parts of it which are dry and above the water also get wet and heavier; gravity then pulls these parts down. This combined effect of capillarity and gravity eventually straightens out the crumbled cloth. You can see this effect vividly by crumpling a piece of coloured tissue paper and dropping it into a bucket of water. You will be able to see the colour of the tissue darken as water soaks into it and the wet parts lower themselves on to the water surface.

The Invisible Man

In order for the "invisible man" to see, images of external objects must form on the retina of his eyes. This requires refraction of light at the outer surface of his eyes, which cannot occur. Moreover, some light energy must be absorbed by his retina in order for his brain to be triggered into interpreting the image. But then since his eyes would become visible to others, an invisible man must necessarily be a blind man! H.G.Well's "invisible man" could, however, see. This is scientifically impossible.

Hiccupping Charlie

When we eat or drink, a valve at the top of the wind-pipe (which goes from the throat to the lungs) closes in order to prevent food or drink from choking it. This closing is triggered by a signal from the brain indicating that we are about to swallow or drink. Another signal triggers the opening of this valve when we want to breathe in. At the same time, a diaphragm in the chest cavity is pulled downwards, thereby sucking air into the lungs. When we are distracted while eating or drinking, two opposite signals may go to the muscles of the diaphragm and those controlling the valve, setting them off to work against each other. This gives a jerk to the diaphragm, resulting in the hiccup. It is for this reason that one should never pour any fluid into the mouth of a person who has fainted, since the fluid may enter the wind pipe choking the person to death.

The Humming Wires

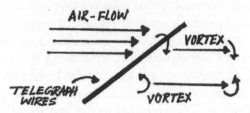

You might think that the vibrations of the telegraph wires in the wind produce the humming sound. Although these vibrations do produce some noise, they are not the main factors. When a fairly high speed wind hits a telegraph wire, the air flow becomes turbulent. Above a certain critical speed, two symmetrically placed stable vortices develop across the telegraph wire. These vortices become unstable when the speed of the wind crosses an even higher threshold value. Then, if one of the vortices is somehow disturbed, it starts oscillating and ultimately breaks away. This is followed by the formation of other vortices in place of the earlier ones. This is technically known as the "hydrodynamic feedback" mechanism. As a result, a chain of alternating vortices flow away from the telegraph wire. These vortices are accompanied by rapid pressure variations in the surrounding air, which generate the characteristic humming sound. It was Lord Rayleigh who first made a systematic study of such phenomena.

Can Lightning Magnetise a Sword?

MAGNETIC LINES OF FORCE

Lightning is discharge of electricity, and the resulting electric current passing through an iron sword would produce magnetic lines of force coiled around it. What are needed for magnetisation, however, are the lines of force parallel to and along the length of the sword. However strong the current from lightning might be, it cannot magnetise a sword unless it produces magnetic lines of force parallel to the sword. Moreover, even if the sword were somehow to acquire some magnetisation, it should be completely destroyed by the enormous heat generated by the huge current arising from the lightning discharge.

The Ben Hur Chariot Race

In a film projection, 24 frames are projected per second. As a wheel picks up speed, there comes a stage when the wheel looks still. This is known as the "stroboscopic" condition. It is realized every time the wheel speed increases to a point where the configuration of the spokes remains unchanged over successive picture frames. However, just before this condition is realized, the wheel speed is such that the spokes just fail to arrive at their previous configuration. The wheel therefore appears to move backwards in spite of the increase in its speed. As the stroboscopic condition is reached, the wheel comes to rest, and then starts moving forward as it speeds up further.

Doctor Zhivago

When you look at trees in summertime, you see only one colour—green. Of course, there are various shades of green, but it's as if it were all painted by one brush. The leaves have chlorophyll, but they also contain smaller quantities of other pigments, like xanthophyll (which produces yellow colours), anthocyanins (which produce the bright red colours) and carotene (which produces orange). But because of the predominance of chlorophyll, these colours are obscured and the leaves look green most of the time.

With the onset of the cold season and the weakening of sunlight during shortened days many trees shut down their food factories, and the food stored in the leaves flows out to the branches and trunks. The chlorophyll disintegrates and, as it disappears, the other pigments show up—the leaves turn red and gold. Eventually, the dry leaves fall and provide nutrients to the soil below. This is the culmination of the process that begins in spring.

Of course, there are a great many evergreens—trees that go on looking lush and green. They shed their leaves throughout the year. These either occur in areas where sunlight is more or less uniform over the year, or they have specially shaped leaves like the conifers.

The Green Flash

This phenomenon involves the absorption of sunlight by water vapour as well as its scattering and refraction in the atmosphere. A major part of the yellow and orange components is absorbed by water vapour while the violet and blue components are scattered away by air molecules. What remain are the red and green components. Since the atmosphere is denser nearer the ground, refraction splits the red and the green because of their different refrangibilities. Red is bent less than the green, resulting in the green upper rim of the sun, the so-called "green flash". As pointed out by Michael Berry in his letter to *New Scientist* (30 November 1991), it can sometimes be seen as a "deep green gleam". Berry mentions that the moon can show the green flash too; he says he has observed with a telescope the green upper rim of the moon when the moon is low in the sky.

The "green flash" is usually clearest on evenings when the sun is dazzling white until very close to the moment of setting; it is never seen when the sun is very red well before setting. Can you figure out the reason?

The book titled *The Green Flash and other Low Sun Phenomena* by D J K O'Connell (North-Holland, 1958) contains spectacular photographs of green flashes. In an article published in *Sky & Telescope* (Volume 12, p 233, July 1953) T S Jacobsen mentions that the higher the latitude, the longer is the period over which the "green flash" can be seen. During Byrd's expedition to the South Pole, the "green flash" was seen for 35 minutes as the Sun rose for the first time at the end of a long polar night and moved along the horizon.

The Murmuring Brook

It is the volume pulsation of trapped air bubbles in the stream that produces the murmur. The pulsating air bubbles behave like oscillating systems (bells) and generate sound waves in the audible range. You can create this murmur at home. Take two glasses partially filled with water. Pour water from one into the other and listen to the murmur. Notice that air bubbles form in the water.

V Fly

When a bird flaps its wings downward, it forces updraughts of air which trail beyond its two wings. Another bird following it is able to take advantage of these updraughts if it positions itself just behind the tip of one of the wings in order to avoid coming in each other's way. It is therefore most advantageous for migratory birds to fly in V formations. In this way they spend the least amount of energy. This is useful for very long distance flights which migratory birds have to undertake for survival. How did these birds learn this trick ? We guess that evolution through natural selection may be the answer. Those species that did not develop the required "instinct" have not survived. Would you agree?

Light and Shade

It is of course obvious that it is sunlight piercing through apertures in the tree canopy that gives rise to these bright patches on the ground. To understand why all of them have the same elliptical shape, it would be instructive to do a simple experiment. Intercept one of these light spots by a piece of paper held at right angles to the rays, and you will see that it is no longer elliptical—it is circular. Raise the paper higher along the rays and the spot grows smaller. This shows that the light rays producing the spot form a cone with its apex on the tree top. The spots on the ground are elliptical simply because the ground cuts these cones slantwise. The origin of this phenomenon lies in the fact that the sun is not a mere point—every small opening in the tree canopy forms an extended image of the sun on the ground.

Wet Bottomed

Under normal atmospheric pressure ice melts at 0°C. However, with increased pressure on it ice melts at a lower temperature. The bottom of a glacier is under considerable pressure due to the weight of the ice above. As a result, the ice at the bottom melts easily. If this didn't happen, the continuous accumulation of ice would have made glaciers eventually collapse under their own weight, resulting in disastrous avalanches.

Raman Confronts Rayleigh

Raman made use of the fact that reflection by a smooth surface polarizes light (in plane polarized light, the electric field oscillates in a fixed plane). For certain angles of observation, the degree of polarization can become very high. A nicol prism has the special property of transmitting only light polarized in a particular plane. It is therefore possible to cut off plane polarized light appreciably by orienting the nicol suitably. This is precisely what Raman did —he looked at the reflected light from the sea through a nicol prism which he happened to carry with him and turned the nicol around to cut off the reflected light. He was surprised to find a beautiful blue light still emerging from the sea. This clearly showed that the blue of the sea could not be entirely due to the reflection of the sky.

Raman followed this up with more thorough investigations and published his findings in the *Proceedings of the Royal Society (London),* Volume 101A, p 64, 1922. This led him to the idea that scattering of light by liquid molecules could be an important factor accounting for the blueness of the sea. He pioneered a series of experiments on the scattering of light by various liquids which ultimately won him the Nobel Prize in 1930 for discovering a totally new and unexpected effect named after him. It all started with that simple experiment on board a ship!

Shades of Blue and Green

There are several factors that determine the colour of the sea :(a) Part of the light falling on the sea is reflected by the surface. This reflected light is predominantly blue or grey depending on whether the sky is clear or clouded. Reflection is a significant factor determining the colour of the sea but is not the only factor. (b) An appreciable part of the light is scattered by the water molecules and this too is blue. There is also some scattering by the particles of sand or clay and this scattered light is mainly brownish in colour. (c) Another important factor is the absorption of light by water molecules which absorb the red, orange and yellow components of the spectrum. The colour of seawater on its own is thus the combined effect of scattering and absorption. (d) The green colour of certain parts of the sea at certain times remained a puzzle until the Indian physicist K R Ramanathan discovered in 1925 (*Philosophical Magazine*, Volume 46, p 543) that certain organic substances (types of plants) in the sea absorb the blue and violet components of white light and emit green light by a process known as fluorescence. (e) Finally, the varying conditions of the sky and the sea water contribute significantly to the changing patterns of the colour of the sea.

The Blue Dome of Air

Light from the sun spreads out through space like ripples on the surface of a pond. These ripples or waves have very small wavelengths of the order of 0.00006 cm. When they fall on air molecules (in the earth's atmosphere) which are much smaller in size, the waves are scattered from air molecules in a particular fashion and these scattered waves reach our eyes giving rise to the colour of the sky. Lord Rayleigh was the first to show theoretically that the intensity of the scattered light should increase sharply as its wavelength decreases.

In the visible spectrum of sunlight violet has the shortest wavelength. It, therefore, follows that more violet light should be scattered into our eyes than blue, green or red light. But then why does the sky look blue rather than violet? That is because of two other important reasons. First, there is more blue light in the rays coming from the Sun than violet. Secondly, our eyes are much less sensitive to violet than to blue light (through evolution the human eye is adapted to be most sensitive to the colour most abundant in sun's rays, which happens to be yellow). These factors make the resultant visual sensation dominantly blue.

Why Peaks Peak

If a mountain goes too high, it shrinks into the earth because the material comprising the earth's surface and the rocks at the base—the granite, quartz or silicon dioxide—cannot hold its weight. There is a limit beyond which a solid begins to yield when the bonds between the atoms lose their directionality. In the words of the eminent physicist Victor Weisskopf : ". . .the whole bonds between the atoms are not broken, just the directionality of the bonds. This enables a liquid to flow, whereas a solid cannot because its bonds are held in fixed positions relative to the constituent atoms" (lecture given at CERN, Geneva, 1967, in which Weisskopf discussed this problem). The energy necessary to break the directionality of the bonds, that is to liquefy, comes from the potential energy lost by the mountain as it sinks. Weisskopf did quantitative estimates which show that a mountain on earth cannot be higher than about 30 kilometres. A further reduction in height to about 10 km (the height of Mount Everest !) occurs because the mantle which supports the earth's crust is not rigid—it is plastic and the mountains float on it like gigantic icebergs. Geologists and physicists working together have found that every mountain indeed has an "inverse mountain" (the portion of it submerged in the mantle). This is why a plumb line near a mountain is not deflected towards it as much as one would expect had all the matter making up the mountain been contained in the visible volume.

At first this was a very surprising discovery which made scientists suspect that most mountains were hollow. A proper understanding came with the discovery of the "inverse mountain".

On other planets this critical height would be different because the strength of gravity is different and the planets may also be made of different types of material.

Frozen Over
Ice has a lower density than water and hence floats in water. Being a bad conductor of heat, it insulates the water below and keeps it above the freezing point. This is a blessing since otherwise the whole mass of water from top to bottom would have solidified, destroying all aquatic life. Moreover, even when the temperature rises slightly above 0°C, the top layer of the ice does not melt immediately. This is because ice can sometimes remain in a metastable solid state even above its melting point. This depends on its state of purity.

In Dribs and Drabs
This is essentially because of the growth of water droplets in clouds. A cloud begins to form when vapour condenses around minute dust and other electrically charged particles. These droplets have diameters typically in the range of 1 to 10 microns

(1 micron is 10^{-4}cm). Clouds have significant vertical velocities (typically 1 to 10 m/s) and the water droplets in them are carried upwards. As the droplets move up, they grow bigger due to additional condensation. Larger drops are formed when two or more droplets collide and merge. The moment these droplets grow so large that air can no longer carry them upwards (that is, when the weight of the droplets exceeds the drag force exerted on them by the rising air), they start to fall down as rain. This is a continuous process and this is why we do not see a cloud coming down as rain all at once, except in rare "cloud bursts".

Footsteps on Sand

When the surface of a sandy beach is not too wet, the sand grains are packed as closely as possible. When one steps on it, the grains are squeezed and they rearrange themselves. In the new arrangement the volume increases and hence the pore spaces between the sand grains increase. Consequently, the water flows down to occupy these newly created spaces, making the footprint look dry compared to the rest of the sand.

Murmur in Sea-Shells

A sea-shell acts as a resonator, like the resonators of musical instruments. However, there is a difference. The shell amplifies the low noise in the surroundings which would otherwise remain inaudible, even the noise of the breeze that flows around us. Being a sea-shell, it reminds people of the murmur of the sea. Actually, you can use any cup, including your own palm, to hear the same sound.

Migratory Birds

A complete understanding of how migratory birds find their way across enormous distances is still lacking. However, a vartiety of features have been discovered through ingenious experiments. One of these (designed by Gustav Kramer) involved putting a few migratory birds in a specially designed dark cage with windows and mirrors which allowed the direction of sunlight entering it to be changed. It was found that the birds always took their cue from the direction of the rising sun—they changed their flight direction as the direction of the sunlight was changed. As the day progresses, the birds are able to maintain this direction with the help of a clock within their body.

In another experiment, some migratory birds were put inside a planetarium to simulate early evening conditions. It was found that as the star positions were changed on the dome, the birds changed their direction accordingly.

Experiments have been done with migratory birds grown from the artificial hatching of eggs in incubators. Surprisingly, these birds too show migratory properties which indicate their genetic origin—the relevant information about the flight routes is coded in their genes.

It has also been observed that with the onset of the season for migration these birds automatically develop certain physiological changes (such as secretion of certain hormones, growth of excess fat) which make them restless and propel them to migration. Experts believe that, in addition, many other factors such as the direction of the earth's magnetic field, the earth's daily rotation, variations in the barometric pressure, could provide signals to these birds helping them to monitor their flight routes.

White Surf

The surf is made up of innumerable bubbles. A bubble is trapped air within a very thin film of water which reflects light. Since a bubble has very little liquid in it, it also absorbs very little light compared to a water droplet of the same size. This makes the myriads of bubbles in the foam reflect a large amount of sunlight, giving rise to the white surf we see. This also explains why the foamy head on beer (containing bubbles) is white whereas the liquid from which it is formed is yellowish. Bubbles in beer originate from dissolved carbon dioxide gas which is in equilibrium with evaporated gas occupying the space above the beer in a capped bottle. When the bottle is uncapped, the equilibrium is disturbed and the excess gas escapes in the form of small bubbles.

A curious thing is that the white surf is also faintly visible at night. This is because sea water contains phosphorescent materials. But then why does only the surf glow? Any idea?

Eelectricity

The electric eels use a combination of "series" and "parallel connections". The batteries in your torch light or transistor radio are connected in "series"— that is, the positive end of one is connected to the negative end of the next one and so on. In this arrangement, the total voltage obtained is the sum of the individual voltages, but the same current flows through the circuit. One can also arrange the cells in "parallel" by connecting all the positive ends together and all the negative ends together. Then the voltage is the same as that of a single cell but the currents add up.

An electric eel has certain cells in its body

which produce a flow of current across their membranes when triggered by a signal from the brain. A large number (of the order of thousands) of these cells are arranged in "series" and a large number of these series are connected in "parallel". The "series connections" ensure high enough voltage between the head and the tail of the eel while the "parallel connections" result in a high enough current to kill its prey. This helps to keep the current through each cell within a limit so as not to damage the tissues.

A Flash of Lightning

Before a thunderstorm, the clouds that gather over the earth get charged with electricity. When such a cloud—with a negatively charged underside—gathers over the earth, its potential is much more negative than the earth below. So it has a tendency to offload the accumulated charge to the ground, and get rid of its excess energy. But before that can happen, it has to bridge the gap through the insulating air.

It all starts with what is called a "step leader"—a little bright spot that originates from the cloud and moves rapidly down at one-sixth the speed of light! It travels about 50 metres, stops for about fifty microseconds, takes another step, pauses, takes another step . . . and so on—describing a zig-zag path down to the earth.

Since this leader carries negative charges from the cloud, the whole column is suffused with negative electricity.

The moment the leader touches the ground, we have a conducting "wire" that runs all the way up to the cloud. Now negative charges in the cloud can run out, producing positive ions which run up the same path, further ionising the molecules in the air that stand in their way, producing a gigantic flash of light. This is why the lightning stroke we see runs from the ground upward. This stroke—the brightest part of the lightning—is called the return stroke. The heat generated by the flash causes a rapid expansion of air and makes a thunder clap. The sound is repeatedly reflected from cloud to cloud to create the rumbling we hear.

The Chameleon Moon

During the day the moon appears white because the intense blue scattered by the sky is added to the moon's own yellowish light. As the sun sets, the blue light of the sky gradually disappears and the moon appears yellower. Eventually at a certain moment it looks pure yellow. As the night darkens, the moon returns to yellow-white, since the moon becomes brighter than the background. Because of purely psychological reasons, bright sources of light tend to appear white to us. In fact, on a very clear night when the moon is at the zenith, it looks almost pure white.

Brighter than the Sky

Objectively, this is not possible. However, this is an example of the subjectivity of brightness. We compare snow on the ground with surrounding darker objects such as trees or the sky at the horizon which is usually not as bright as the zenith on an overcast day. If you can manage to look at the snow on the ground as well as the zenith simultaneously (for instance, with the help of a mirror placed on the snow), you will clearly see that the zenith is indeed brighter than the snow. A similar thing happens with falling snowflakes—they change abruptly from being black to white as their backdrop changes from a bright sky to dark woods.

The Colour of Smoke

When the background is bright, we are able to see smoke by the light transmitted through it. It appears yellowish red because the violet, blue and green components of white light are scattered more by the particles making up the smoke, and it is mainly the yellow and red components that are able to come through the smoke and reach our eyes. On the other hand, when the background is dark, the smoke is illuminated by light from the sun (or some other source) falling on it from behind us. In this case, it is scattered rays from the smoke which enter our eyes and make the smoke visible. Since the blue part of white light is scattered most by smoke particles, we see them as blue. A beautiful description of this phenomenon was given by J Tyndall: "Years gone by, I used to see something similar to this in Killarney, when on windless days the columns of smoke rose above the roof of the cottages. The lower part of each column was shown up by a dark background of pines, the top part by the light background of clouds. The former was blue because it was mainly seen through dispersed light, the latter was reddish because it was seen through transmitted light".

Dew Point

For dew to form on a surface, the temperature must drop below that of the dew point of the air to which it is exposed (dew point is the temperature at which air is saturated with the water vapour present in it). As water vapour starts condensing, dew forms on the surface. However, at the same time, some of the rapidly moving water molecules in the dew film start escaping by evaporation. The keypoint is that only when this rate is less than the rate of condensation does dew form.

All objects at temperatures above absolute zero emit heat radiation. On a typical clear winter night, objects on the ground emit more heat than they receive from the sky and consequently become cooler. Now, the lower the temperature of the surface, the less is the rate of evaporation of water from it. This accounts for the copious formation of dew on clear nights.

Most polished metal surfaces emit and absorb radiation at a lower rate than insulators under identical conditions. The difference between emission and absorption is also less for metal surfaces and consequently they are cooled less. This is why dew is less likely to form on a polished metal surface.

Winter Veil

The reason lies in what is known as "temperature inversion". There are no strong air currents in the winter to disperse pollutants like smoke either in the vertical or horizontal directions. Also, the ground is not heated very much in the winter. As the sun goes down, the ground radiates heat into a clear sky and cools down fairly quickly. As a result, a layer of cold air gets trapped near the ground below warmer and lighter air above. This is the reverse of the condition that normally prevails (namely, the temperature of air drops with the height above the ground). The cold air near the ground cools all the smoke and traps it below the warmer air above.

The Ghostly Moon

The reason is that when the earth comes between the sun and the moon and casts its shadow on the moon, some sunlight is still refracted by the earth's atmosphere and falls on the shadow region, an effect which is not mentioned in the usual textbook discussions of the lunar eclipse. This refracted sunlight is depleted of its bluer components because of the scattering of light by air molecules (called Rayleigh scattering). Air molecules are smaller than the typical wavelengths of light, and they scatter blue light much more than red light. This is why sodium vapour light is much more effective in lighting streets than blue mercury light. Being yellowish, it is scattered much less than the bluer mercury light and can penetrate deeper. This is also why the light that falls on the moon during a total eclipse is faint and reddish.

Catch a Full Rainbow

The centre of a rainbow is always in line with our eyes with the sun behind us, so that standing on the earth, we are only able to see semi-circular rainbows. The lower halves of the bows are cut off by the earth. A complete rainbow can only be seen when it is formed parallel to the earth's surface, as seen by the passenger in an aeroplane flying above the water droplets with the sun high up above it.

The Moon and the River

The crux of the matter lies in the fact that the moon is a distant object and the height of an aeroplane above the ground is negligible compared with this distance. Now, for an image formed by regular reflection, the image distance is the same as the object distance. The image of the moon in the river is also therefore very distant. Since the height of the plane is negligible compared to this distance, the moon's image will appear to have the same size from the ground and the plane. However, the width of the river will appear to shrink as we go up. Hence there will come a point above which the river will appear narrower than the reflected moon.

Ignorance is Bliss

High voltage of electrical lines is in itself not a problem—what matters is the *voltage drop* between the two points. The voltage drop across the two legs of a bird sitting on a high tension line is fairly small. Coupled with the fact that the electrical resistance of its body is high, this means that practically no current flows through its body. However, if an unlucky bird happens to touch the pole while sitting on a high tension line, there is a short circuit from the electrical line to the earth and a massive current flows through its body, electrocuting it.

Buzzing Bees

Bees and other insects buzz when flying around. The buzzing comes from the flapping of their wings. Whenever anything vibrates more than 16 times a second, it emits a tone of definite pitch. It is this pitch, when matched with a musical note, that tells scientists how many times a second an insect flaps its wings. Do you know that the buzz of an ordinary housefly matches the tone F ? It flaps its wings 352 times a second. Honey bees, when not burdened with honey, flap 440 times a second—they emit the tone A. Loaded with honey their pitch drops to the tone B, which means their wings flap 330 times a second. Even innocent hummings can give us important scientific information!

The Elusive Cricket

Our ears can determine the direction of a sound source in two ways : (a) by noticing the difference in the intensities of the sound heard by the two ears, or (b) by perceiving the difference in the phases of the sound waves reaching the two ears. Both form the basis of stereophonic hearing. Intensity differences are discernible by human ears only for shortwave length or high pitch sound. This is because long wavelength or low pitch sound can diffract or bend round the head and produce equal intensities at the two ears. For such low pitch sound, our ears have to depend solely on their ability to detect the difference of phase at the two ears. At intermediate pitches (~4000Hz) which roughly correspond to the sound produced by a cricket, the location of the source becomes particularly tricky—our ears then find it difficult to differentiate between either the intensities or phases at the two ears.

Pondskater

The surface of water is like a thin stretched membrane or "skin" (surface tension) which can support objects which are not too heavy, nor wetted by water and which do not prick the skin. Insect legs are covered with a web of hairs trapping air which acts like a cushion and are not wetted by water. Their legs simply depress this water "skin" created by surface tension. The skin tends to straighten out and support the insects.

The feather of water birds like ducks are covered with an oily substance exuded by their glands. This is why water does not wet their feathers.

Sap in the Cap

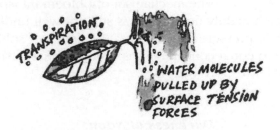

Sap is lifted up to the leaves and then flows down with the products of photosynthesis. Water ascends from the roots through tubes of dead cells in the xylem. Products of photosynthesis descend from the leaves through living cells of the phloem. Experiments have demonstrated that the "motor" of sap ascent lies in the crown of the tree and is powered by sunlight. When the leaves are engaged in photosynthesis, they liberate copious quantities of water vapour to the air, a process called "transpiration". As water transpires, a molecule at a time, from the pores on the under-surface of the leaves, they are replaced by molecules pulled up from below by surface tension forces. The water column is continuous all the way from the rootlets to the capillaries in the leaves. It is therefore not the atmospheric pressure that is utilised but the cohesive forces within water and the adhesive forces between water and cell walls. These cohesive and adhesive forces give a continuous water column a tensile strength as high as 300 atmospheres. The formation of a single air bubble can, however, ruin this mechanism and make the sap drop to approximately 33 feet. That such a delicate mechanism can work reliably in the high, wind-tossed branches of a tree is because of the minute subdivision of the chambered structure of the wood. If a gas bubble forms in a column, the resulting break is confined to that column alone.

The mechanism of phloem transport, although mainly downward, is still not well understood. Osmotic pressure (the universal tendency of solutes to come to equal concentrations everywhere in a solution) could be responsible.

Darkness at Noon

The answer lies in the fact that water is closer to sand than air in its optical properties. Light is scattered by the sand grains but emerges fairly quickly after a few scattering events because the average scattering angle is large. When the inter-particle spaces are filled with water (even if it is pure) the average scattering angle is smaller and light suffers a larger number of scattering events and has to travel a longer distance within the sand before re-emerging. It is this longer path and the consequent cumulative absorption by the scattering centres (sand grains) that make wet sand look darker. It has very little to do with absorption by impurities in water. To convince yourself, use washed and clean sand as well as distilled water—the wet sand will still look darker.

The Shape of Ripples

You might think that the waves will take some kind of an elliptic or oblong shape, somewhat wider along the direction of the stream. This is not true. The shape will remain circular in flowing water. The reason is this: the flow will translate the entire body of water downstream. Consequently, the circular waves will undergo a simple translation downstream without suffering any distortion.

Twinkle, Twinkle, Little Star

The twinkling of stars is caused by earth's atmosphere. If you were on the moon, for example, where there is no atmosphere, you would not see stars twinkle.

Stars are so far away that they act as point sources of light. Due to constant air currents in the atmosphere, the density fluctuates. As a result, the light rays from a star undergo random deviations (refractions) as they pass through the atmosphere. These density fluctuations vary in time resulting in rapid changes in the apparent position of a star. But what exactly makes these quick variations of the perceived position give rise to what is known as the twinkling effect is not entirely clear. This could well involve subtle aspects of the intricate mechanism of our visual perception.

Planets are comparatively nearer to the

earth and they look like small discs of light rather than point sources. Although the turbulence in the earth's atmosphere produces fluctuations of each point in the disc, these fluctuations more-or-less cancel each other out over the disc and the average effect is one of steady light.

The twinkling of a planet may become noticeable when light rays from different points on its surface suffer large deflections that shift their images outside the planet's "apparent" diameter. This can happen if the disturbance in the air is very pronounced and the planets are low in the sky, as is the case with Venus and Mercury at times.

Isaac Newton (1642–1727) published his book Opticks *in 1704 in which he explained the rainbow and proposed the "corpuscular" (particle) theory of light. In his book* Mathematical Principles of Natural Philosophy *he set down the principles of classical mechanics and gravity.*

Finally, a historical sidelight : Newton had briefly commented in his book *Opticks* (4th edition, p 110) on why the stars appear to twinkle. He put it in this way: "For the air through which we look upon the stars is in a perpetual tremor; as may be seen by the tremulous motion of shadows cast from high towers and by the twinkling of a fixed star".

The Blue Zenith

The enhanced blueness of the zenith is owing to the presence of ozone in the upper layers of the atmosphere. Absorption of light by ozone is highest at the red end of the spectrum and is least at the blue end. When the sun is just below the horizon, the path length of sunlight through the ozone layer is the greatest for the light scattered from the zenith, and consequently it is most depleted of its reddish components.

Once in a Blue Moon

What Robert Wilson concluded from his observations was that the blue sun and moon were caused by clouds of small particles from forest fires in Alberta (Canada), which had been carried by winds across the Atlantic to Edinburgh. These particles were predominantly oil droplets formed from the combustion products of the fires. The oil drops had sizes comparable with the average wavelength of light. Now, we know that if the scattering particles are much smaller than this, they preferentially scatter blue light (Rayleigh scattering). If they are much bigger, they scatter all colours more or less equally. When they are of a comparable size, they scatter red light more than blue. It so happened that the oil drops carried by the Canadian forest fires to Edinburgh were just the right size to scatter away red light more than blue. Consequently, the Sun and moonlight that got through and reached our eyes looked blue. It was indeed a very rare combination of factors. Such a combination occurs only once in a blue moon!

Halo Moon

The halo around the moon is caused by refraction and dispersion of light. There are thin white clouds in the sky, so thin that we can see the moon through them. These clouds are made up of tiny hexagonal ice crystals. Sun's rays, reflected by the moon, while coming through the crystals are refracted (as in a prism). The refraction is accompanied by dispersion, that is, splitting into colours. The halo appears pinkish because of the central pink colour, which can be seen distinctly, while the outer blue colour merges into the background of the sky. The halo looks circular which implies that the ice crystals are uniformly distributed around the centre of the halo.

Olbers' Paradox

Astronomical observations do not so far indicate any bounds on the spatial extent of the universe. There is also strong empirical evidence that the distribution of luminous objects in the universe is remarkably uniform. Assumptions (a) and (b) are, therefore, unlikely to be false. One might suspect the assumption (e) to be false (as Olbers himself did) because it is conceivable that the absorption of light by intergalactic matter plays a significant role. However, this will not help. While absorption would reduce the light from a distant source, it would at the same time heat up the absorbing material which would then emit its own radiation.

It turns out that assumptions (c) and (d) are really the vulnerable ones. Observations have shown that the universe is not static—it is ever expanding with galaxies receding from one another. This is a theoretical consequence of Einstein's relativistic cosmology. In an Einsteinian universe, a diminution of light reaching us from distant galaxies can occur in two ways. First, there is a difference in the time scales which operate on the earth and on distant and massive galaxies. According to Einstein's general theory of relativity, the geometry of space is affected by the presence of a massive body and this, in turn, affects local time —atomic clocks in massive galaxies appear to run slower than the atomic clocks on earth. This results in shifts of spectral lines towards the red end of the spectrum in visible light coming from massive galaxies. This is known as the "gravitational red-shift". In addition, there is a "cosmological red-shift" (a Doppler shift) in the light from receding galaxies in an expanding universe. Both these effects contribute to a decrease in the energy content of the radiation. This is because of the quantum nature of radiation—light is made up of photons, the energy of each photon being proportional to the frequency which decreases when a "red-shift" occurs.

It is now widely believed in the scientific community that the universe must have had a beginning, that is, the universe is not infinitely old. If we assume that the age of the universe is T, then the light reaching us now could not have originated from beyond a distance cT (c is the velocity of light). Thus, a finite age universe implies a limit on the distance of the sources contributing to the brightness of the sky.

All these effects combine to ensure that the distant stars and galaxies in the universe do not make the night sky bright. The relative importance of these effects continues to be hotly debated by the experts. In fine, the answer to "Olbers' paradox" is that the sky is dark at night because there is an "edge" to the universe in time and because the universe is expanding.

In an instructive historical account of "Olbers' paradox" John Gribbin, in his book *In Search of the Big Bang* (Corgi, 1987), mentions that this puzzle was discussed by Edmond Halley well before Olbers. Halley presented a paper on this topic to the Royal Society in 1721 with Newton presiding over the session. Though Newton himself believed in a universe of finite age, surprisingly he failed to realise that this could be used to solve the "paradox", at least partially. To err is human!

Once upon a time an astronomer, a physicist, and a mathematician set off on a walking tour in the Scottish highlands. They soon came across a sheep grazing all alone on a farm. Looking at it, the astronomer commented "So, the sheep in the highlands are black".

"You cannot generalize so sweepingly", admonished the physicist, "Your sample is too small. Only after a careful analysis of a large number of sheep all over the highlands can you make such a statement. Just now all you can say is that black sheep are found in Scotland". He turned to the mathematician for his views.

"I am afraid I disagree with you both", remarked that worthy. "All you can say is that the animal over there appears to be black on the side facing us".

Index

T - #0558 - 101024 - C0 - 216/138/10 - PB - 9780750302753 - Gloss Lamination